# Computational Intelligence in Bioprinting

**Scrivener Publishing**
100 Cummings Center, Suite 541J
Beverly, MA 01915-6106

*Publishers at Scrivener*
Martin Scrivener (martin@scrivenerpublishing.com)
Phillip Carmical (pcarmical@scrivenerpublishing.com)

# Computational Intelligence in Bioprinting


Edited by

**E. Gangadevi**

*Department of Computer Science, Loyola College, Chennai, India*

**M. Lawanya Shri**

*School of Information Technology and Engineering, VIT, Vellore, India*

**Rajesh Kumar Dhanaraj**

*Symbiosis Institute of Computer Studies and Research, Pune, India*

and

**Balamurugan Balusamy**

*Shiv Nadar University, Delhi-NCR Campus, Noida, India*



Scrivener Publishing

WILEY

This edition first published 2024 by John Wiley & Sons, Inc., 111 River Street, Hoboken, NJ 07030, USA and Scrivener Publishing LLC, 100 Cummings Center, Suite 541J, Beverly, MA 01915, USA

For more information about Scrivener publications please visit www.scrivenerpublishing.com.



**Wiley Global Headquarters**
111 River Street, Hoboken, NJ 07030, USA

For details of our global editorial offices, customer services, and more information about Wiley products visit us at www.wiley.com.



***Library of Congress Cataloging-in-Publication Data***

ISBN 978-1-394-20439-7

Cover image: Pixabay.Com
Cover design by Russell Richardson

Set in size of 11pt and Minion Pro by Manila Typesetting Company, Makati, Philippines

Printed in the USA

10 9 8 7 6 5 4 3 2 1

# Contents

# Preface

Bioprinting is an evolving field in reformative medicine. It is well-defined as the printing of structures using feasible cells, biomaterials, and living molecules. The ultimate intention of bioprinting is to offer a substitute for autologous and allogeneic tissue implants, and also to transform animal testing for the learning and analysis of disease and growth of treatments. 3D bioprinting is an innovative skill that will ultimately make medical care faster, more applicable, and more adapted. Normally, 3D bioprinting can employ a layer-by-layer process to accumulate materials called "bio-inks". Bio-inks are an important element of bioprinting and entail biomaterials that can be used to summarize cells and integrate biomolecules. They create tissue-like structures used in many tissue-related medical fields. Bioprinting organs and tissue spots from a patient's cells can decrease the risk of refusal and can eliminate the requirement for organ donors.

This book explores the intersection of computational intelligence and bioprinting, and showcases the potential and current progressions in this rapidly developing field. By harnessing the power of computational intelligence techniques, such as artificial intelligence, machine learning, optimization algorithms, and data analytics, we can overcome the existing hurdles and unlock the full potential of bioprinting.

The chapters herein are subsidized by leading experts and researchers in the arena, who provide complete insights into the challenges and future directions of computational intelligence in bioprinting. The book covers an extensive range of topics, including bio-ink formulation and characterization, bioprinter hardware and software design, tissue and organ modeling, image analysis, process optimization, and quality control. By delving into these key areas, we aim to provide a holistic view of the field and inspire new avenues of research.

We hope that this book serves as a valuable resource for researchers, practitioners, and students interested in the exciting field of bioprinting and its integration with computational intelligence. Whether you are a seasoned expert or new to the field, we believe that the contents presented here will stimulate your curiosity, spark innovative ideas, and drive the next wave of advancements in bioprinting.

We wish to express our sincere gratitude to all the dedicated, passionate authors who contributed their knowledge and insights. We also want to thank the reviewers for their valuable feedback and suggestions, which have significantly improved the quality throughout, as well as Wiley and Scrivener Publishing for their cooperation and assistance in the publication of this book.

Finally, we extend our gratitude to the readers for their interest in this book. We hope that the knowledge shared here will inspire you to contribute to the wonderful field of computational intelligence in bioprinting and jointly work toward converting healthcare and regenerative medicine.

**The Editors**
November 2023

# 1

# The Emergence of Bioprinting and Computational Intelligence

**P.M. Kavitha[1*], S. Jayachandran[1] and M. Anitha[2]**

[1]*Department of Computer Applications, SRM Institute of Science and Technology, Ramapuram, Chennai, India*
[2]*Department of Computer Science and Engineering, SRM TRP Engineering College, Trichy, India*

## *Abstract*

Bioprinting is a promising technology that involves the creation of living tissues and organs through 3D printing techniques. However, the complexity of the structures involved in bioprinting makes it challenging to create viable tissues and organs. Computational intelligence, which enables computers to learn, reason, and make decisions similar to humans, has emerged as a critical tool in the development of bioprinting. Through the use of computational intelligence, researchers can simulate the behavior of cells and tissues in different environments. These simulations can help develop more accurate models for bioprinting and optimize the printing process for the creation of functional tissues and organs. Additionally, computational intelligence can aid in the analysis of data obtained from experiments and simulations, which can be used to refine and improve the bioprinting process. The emergence of bioprinting and computational intelligence has the potential to revolutionize the field of regenerative medicine, allowing for the creation of replacement tissues and organs for patients in need. As the technology continues to evolve, the use of computational intelligence will play an increasingly important role in the development of new bioprinting techniques and the advancement of regenerative medicine. One of the key challenges of bioprinting is the complexity of the structures involved. Unlike traditional 3D printing, bioprinting requires the printing of living cells, which can be highly sensitive to their environment. Computational intelligence can help address this challenge by allowing scientists to simulate the behavior of cells and tissues in different environments. By analyzing data from these simulations, researchers can develop more accurate models.

**Corresponding author*: pmkavimca@gmail.com

E. Gangadevi, M. Lawanya Shri, Rajesh Kumar Dhanaraj and Balamurugan Balusamy (eds.) Computational Intelligence in Bioprinting, (1–20) © 2024 Scrivener Publishing LLC

***Keywords:*** Simulations, 3D printing, regenerative medicine

## 1.1 Introduction

The enthralling era of computational intelligence and bioprinting technologies decides the future of the health care and biological world. Bioprinting deals with the high dimensional printing of the biomedical products like cells, tissues and even organs in a controlled environment, with high accuracy. The goal of bioprinting is to create functional biological structures that can be used for a wide range of applications, such as tissue engineering, regenerative medicine, and drug discovery. The process of bioprinting involves using a printer to deposit layer upon layer of biological materials to build up a 3D structure. This process is similar to traditional 3D printing, but with one crucial difference: the materials being printed are living cells, not plastic or metal.

Computational intelligence, on the other hand, is a subfield of artificial intelligence that focuses on the development of algorithms and mathematical models to perform tasks that would normally require human intelligence, such as learning and pattern recognition. In the context of bioprinting, computational intelligence is used to optimize and control the printing process, allowing for the creation of highly precise and accurate biological structures. So why are bioprinting and computational intelligence such a big deal? Well, the potential applications of this technology are truly staggering. Bioprinting has the potential to revolutionize the way that biological and medical problems are approached. For example, bioprinted tissues could be used for drug testing and development, allowing for more accurate and efficient testing of new drugs before they are tested on human subjects. Additionally, bioprinter organs could one day be used to replace damaged or diseased organs, eliminating the need for organ donors and reducing the wait time for transplant patients. And that is just the tip of the iceberg! As our understanding of bioprinting and computational intelligence continues to grow, it ensures to see even more incredible applications of this technology in the future.

The field of bioprinting and computational intelligence is still in its infancy, but it is advancing rapidly. Just over two decades ago, the first proof-of-concept studies demonstrated the feasibility of 3D printing biological materials. Today, there are numerous commercial companies and academic institutions exploring the potential of this field for a wide range of applications.

An introduction to the exciting world of bioprinting and computational intelligence. This study is helpful to learn about the current state of this

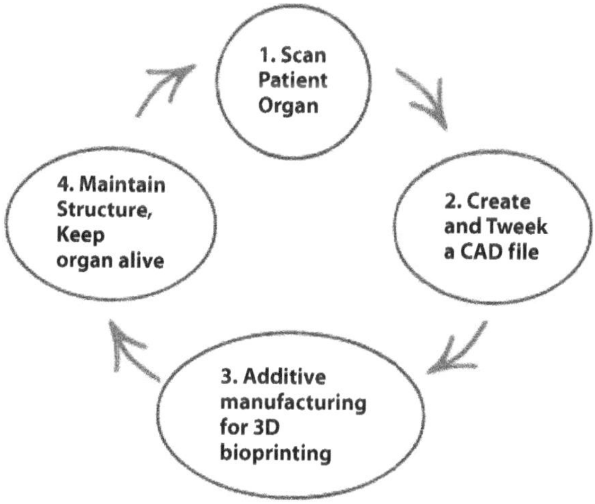


**Figure 1.1** 3D bioprinting—the process.

field, including the key technologies and applications, as well as the current challenges and opportunities. This chapter enables the user to have a comprehensive understanding of the emergence of bioprinting and computational intelligence as a field of study and its potential to shape the future of medicine and biology (Figure 1.1).

The chapter is organized with related study section followed by the Basics of Bioprinting and Computational Intelligence section. Section 1.4 includes The Role of Computational Intelligence in Bioprinting.

## 1.2 Related Study

The article [1] "3D bioprinting: A review on its advancements and future prospects" provides a comprehensive overview of the current state of 3D bioprinting technology and its potential for future applications in biomedical research and clinical practice. The authors discuss the various bioprinting techniques, bioinks, and cell sources currently used in 3D bioprinting, as well as the challenges and limitations of the technology. The article also covers the wide range of tissue types and organs that have been successfully printed using 3D bioprinting, including bone, cartilage, skin, liver, and heart tissue. In addition, the authors describe the emerging areas of research in 3D bioprinting, such as the use of stem cells and bioprinting of vascularized tissues. Overall, the article provides a valuable resource for researchers and clinicians interested in the field of 3D bioprinting, and highlights the potential for this technology to revolutionize regenerative medicine and personalized healthcare.

The article [2] "Bioinks for 3D bioprinting: an overview" provides a detailed review of the various types of bioinks used in 3D bioprinting. The authors cover the most commonly used bioink materials, including natural polymers, synthetic polymers, and hydrogels, and discuss the advantages and limitations of each material. The article also highlights the importance of designing bioinks that mimic the extracellular matrix (ECM) of the target tissue, and provides insight into the strategies used to achieve this goal. The authors also discuss the challenges and opportunities associated with bioink development, including the need for biocompatibility, printability, and appropriate mechanical and biological properties. In addition, the article covers the emerging trends in bioink development, such as the use of decellularized ECMs and the incorporation of bioactive molecules and growth factors. The authors also address the importance of standardizing bioink characterization and testing procedures to ensure reproducibility and comparability across studies. Overall, the article provides a comprehensive overview of bioinks for 3D bioprinting, and serves as a valuable resource for researchers working in this field. The article's focus on the need for tailored bioink development to specific tissue types underscores the importance of ongoing research in this area, and highlights the potential for bioinks to play a key role in the development of regenerative medicine and tissue engineering applications.

The article [3] "Bioprinting of human tissues: current state-of-the-art, challenges, and opportunities" provides a comprehensive review of the current state-of-the-art in bioprinting technology, as well as the challenges and opportunities associated with this rapidly evolving field. The authors cover the various bioprinting techniques and materials used to print human tissues, and discuss the advantages and limitations of each approach. The article also covers the major challenges facing bioprinting technology, including the need for improved biomaterials, the difficulty of vascularizing printed tissues, and the need for better methods of cell sourcing and differentiation. The authors also address the ethical considerations associated with bioprinting, including the need to balance scientific progress with safety and ethical concerns. In addition, the article highlights the opportunities presented by bioprinting technology, including the potential for personalized medicine, disease modeling, and drug screening. The authors discuss the importance of collaborative efforts between researchers, clinicians, and industry partners to drive progress in this field.

Overall, the article provides a valuable resource for researchers, clinicians, and industry partners interested in the field of bioprinting, and underscores the potential for this technology to revolutionize regenerative medicine and tissue engineering. The authors' focus on the importance of addressing the major challenges facing bioprinting technology highlights the need for ongoing research and development efforts in this area.

The article "Printing the future: challenges and opportunities for 3D bioprinting of tissues and organs" provides an insightful overview of the current state of 3D bioprinting technology and its potential for future applications in biomedical research and clinical practice. The authors discuss the various bioprinting techniques, bioinks, and cell sources currently used in 3D bioprinting, and the challenges and limitations of the technology. The article also covers the range of tissue types and organs that have been successfully printed using 3D bioprinting, including skin, heart, liver, and bone tissue. The authors discuss the importance of designing bioprinted tissues and organs that mimic the native tissue architecture, composition, and mechanical properties. They also highlight the need for developing effective vascularization strategies for bioprinted tissues and organs. In addition, the article covers the emerging trends in bioprinting technology, including the use of induced pluripotent stem cells and gene editing techniques to generate cells and tissues for bioprinting. The authors also discuss the potential for bioprinting to transform personalized medicine, disease modeling, and drug screening. Overall, the article provides a comprehensive overview of 3D bioprinting technology, and serves as a valuable resource for researchers and clinicians interested in this rapidly evolving field. The authors' emphasis on the importance of addressing the major challenges facing 3D bioprinting technology underscores the need for ongoing research and development efforts to realize the full potential of this technology for regenerative medicine and tissue engineering.

The article [5] "Advances in 3D bioprinting technology for tissue engineering and regenerative medicine" provides a comprehensive review of the current state-of-the-art in 3D bioprinting technology and its applications in tissue engineering and regenerative medicine. The authors cover the various bioprinting techniques, bioinks, and cell sources used in 3D bioprinting, and the challenges and limitations of the technology. The article also covers the range of tissues and organs that have been successfully bioprinted, including bone, cartilage, skin, liver, and heart tissue. The authors discuss the importance of developing bioprinted tissues and organs with appropriate mechanical properties, cellular architecture, and functionality. In addition, the article highlights the emerging trends in 3D bioprinting technology, including the use of advanced biomaterials, microfluidics, and artificial intelligence to enhance bioprinting precision and control. The authors also discuss the potential of 3D bioprinting for personalized medicine, disease modeling, and drug screening. Overall, the article provides a valuable resource for researchers and clinicians interested in the field of 3D bioprinting, and underscores the potential of this technology to transform regenerative medicine and tissue engineering. The authors' focus on the importance of developing bioprinted tissues and organs with appropriate mechanical properties and cellular architecture underscores the need for ongoing research and development efforts in this area.

## 1.3 Understanding the Basics of Bioprinting and Computational Intelligence

In order to fully understand the intersection of bioprinting and computational intelligence, it is important to first have a good grasp of the basics of each technology. This chapter will provide an overview of the fundamentals of bioprinting and computational intelligence, including a discussion of the key concepts, technologies, and applications.

### 1.3.1 Bioprinting: The Basics

At its core, bioprinting is a process of 3D printing biological materials, such as cells, tissues, and organs, in a controlled and precise manner. The goal of bioprinting is to create functional biological structures that can be used for a wide range of applications, such as tissue engineering, regenerative medicine, and drug discovery.

The process of bioprinting typically begins with the collection of cells from a living organism. These cells are then combined with a hydrogel or other supporting material to create a bioink. The bioink is then loaded into a bioprinter, which deposits layer upon layer of the bioink to build up a 3D structure. There are several different types of bioprinters available, each with their own strengths and limitations. Some bioprinters use extrusion-based printing, in which the bioink is pushed through a small nozzle and deposited layer by layer. Other bioprinters use laser-based printing, in which the bioink is solidified using a laser. Still others use inkjet-based printing, in which the bioink is deposited using a series of small nozzles.

One of the key challenges of bioprinting is ensuring that the printed cells remain viable throughout the printing process and after printing is complete. This requires a delicate balance between the mechanical stress of the printing process and the needs of the cells, such as access to oxygen and nutrients.

### 1.3.2 Computational Intelligence: The Basics

Computational intelligence is a branch of artificial intelligence that focuses on the development of algorithms and mathematical models to perform tasks that would normally require human intelligence, such as learning and pattern recognition. In the context of bioprinting, computational intelligence is used to optimize and control the printing process, allowing for the creation of highly precise and accurate biological structures.

There are several key components to computational intelligence, including machine learning, evolutionary algorithms, and fuzzy logic. Machine learning algorithms use data and statistical models to enable systems to learn and improve over time, without being explicitly programmed. Evolutionary algorithms use principles from natural selection and genetics to optimize systems, such as finding the best possible solution to a problem. Fuzzy logic is a form of mathematical logic that allows for uncertainty and vagueness in decision-making processes.

In bioprinting, computational intelligence is used to control the printing process and ensure that the printed structures are accurate and consistent. For example, machine learning algorithms can be used to optimize the printing process, such as controlling the speed and temperature of the printer to ensure that the printed structures are of high quality. Evolutionary algorithms can be used to find the best possible solution for a specific bioprinting problem, such as optimizing the composition of the bioink. And fuzzy logic can be used to make decisions about the printing process, such as adjusting the printer's settings based on the conditions in the surrounding environment.

### 1.3.3 Applications of Bioprinting and Computational Intelligence

The combination of bioprinting and computational intelligence has the potential to revolutionize the way that is approached biological and medical problems. Some of the key applications Figure 1.2 of this technology include:

- Tissue engineering: Bioprinted tissues can be used for drug testing and development, allowing for more accurate and efficient testing of new

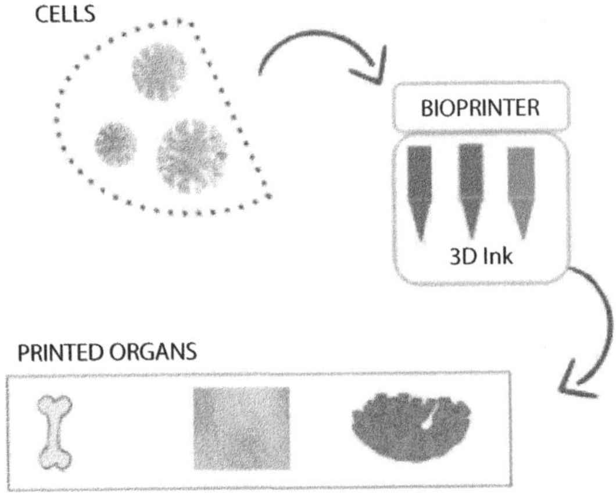


**Figure 1.2** Applications of bioprinting.

## 1.4 The Role of Computational Intelligence in Bioprinting

Computational intelligence plays a critical role in bioprinting, allowing for the creation of highly precise and accurate biological structures. By using algorithms and mathematical models to control the printing process, computational intelligence enables bioprinters to produce structures that are not only functional, but also consistent and repeatable. One of the key applications of computational intelligence in bioprinting is in the optimization of the printing process itself. For example, machine learning algorithms can be used to analyse data from previous bioprinting experiments to identify patterns and make predictions about the outcome of future prints. This information can then be used to make adjustments to the printing process, such as adjusting the temperature or speed of the printer, to improve the quality and accuracy of the printed structures.

The key role of computational intelligence in bioprinting is in the design of the bioink. Evolutionary algorithms can be used to optimize the composition of the bioink, taking into account factors such as the viability of the cells and the mechanical properties of the supporting material. This allows for the creation of bioinks that are better suited for the specific requirements of each bioprinting project. Computational intelligence can also be used to monitor and control the printing process in real-time, ensuring that the printed structures remain accurate and consistent. For example, fuzzy logic algorithms can be used to adjust the settings of the bioprinter based on the conditions in the surrounding environment, such as temperature and humidity. This helps to ensure that the printed structures are of high quality and that the cells remain viable throughout the printing process.

Finally, computational intelligence can be used to analyze and interpret the data generated by bioprinting experiments. This can include data on the viability of the cells, the mechanical properties of the printed structures, and the response of the cells to different treatments and conditions. By using machine learning algorithms to analyze this data, researchers can gain a deeper understanding of the underlying biological processes and identify new and innovative approaches for improving bioprinting technology.

In conclusion, computational intelligence plays a crucial role in the field of bioprinting, enabling the creation of precise, accurate, and functional biological structures. Through the use of machine learning, evolutionary algorithms, and fuzzy logic, computational intelligence allows bioprinters to optimize and control the printing process, leading to better outcomes and more successful bioprinting experiments.

## 1.5 Applications of Bioprinting and Computational Intelligence in Medicine

Bioprinting and computational intelligence are rapidly becoming important tools in the field of medicine, with a wide range of applications that are transforming the way healthcare is approached. From the creation of customized prosthetics to the development of new treatments for diseases, bioprinting and computational intelligence are revolutionizing the way that medicine is practiced.

One of the most promising applications of bioprinting and computational intelligence in medicine is in the field of tissue engineering. Bioprinters can be used to create functional, three-dimensional structures that can be used to repair or replace damaged tissues in the body. For example, bioprinted skin and bone can be used to treat injuries and congenital conditions, while bioprinted liver and heart tissue can be used to model diseases and test new treatments. Bioprinting and computational intelligence are also being used to create customized prosthetics and implants that are tailored to the unique needs of each patient. By using computational algorithms to analyze data on the patient's anatomy, bioprinters can create implants that are perfectly fitted to the patient's body, improving the comfort, functionality, and long-term outcomes of these devices.

In addition to tissue engineering and prosthetics, bioprinting and computational intelligence are also being used to develop new treatments for diseases. For example, bioprinters can be used to create models of diseased organs, such as the liver or heart, which can then be used to test new drugs and therapies. This allows researchers to identify the most effective treatments for a wide range of conditions, from cancer to heart disease, and to develop personalized medicine plans for individual patients. The Chief aspect of bioprinting and computational intelligence in medicine is in the field of regenerative medicine. By using bioprinting technology to create functional tissues and organs, researchers are exploring new and innovative ways to repair and regenerate damaged tissues in the body. This has the potential to revolutionize the treatment of conditions such as spinal cord injuries, traumatic brain injuries, and even certain types of blindness.

Finally, bioprinting and computational intelligence are also being used to improve the accuracy and precision of medical procedures, such as surgeries and biopsies. For example, bioprinters can be used to create 3D models of patient anatomy, which can then be used to plan and simulate surgeries in advance. This allows surgeons to better understand the patient's anatomy and to identify any potential complications before the procedure is performed, leading to safer, more effective surgeries. Thus, bioprinting and computational intelligence are rapidly transforming the field of medicine, with a wide range of applications that are improving the quality of life for patients and advancing the practice of medicine. From tissue engineering and prosthetics, to new treatments for diseases and improved surgical procedures, bioprinting and computational intelligence are providing new and innovative solutions to some of the most challenging problems in healthcare today.

## 1.6 Bioprinting and Computational Intelligence in Tissue Engineering and Regenerative Medicine

Tissue engineering and regenerative medicine are areas of medicine that are rapidly being transformed by bioprinting and computational intelligence. By leveraging the power of these technologies, researchers are exploring new and innovative ways to repair and regenerate damaged tissues in the body, leading to new and effective treatments for a wide range of conditions. Bioprinting technology is being used to create functional, three-dimensional structures that can be used to replace or repair damaged tissues in the body. By using a combination of cells, growth factors, and biomaterials, bioprinters can create structures that closely mimic the function and structure of native tissues. This is particularly important in the field of regenerative medicine, where the goal is to repair or regenerate damaged tissues in the body.

Computational intelligence, on the other hand, is being used to optimize the design and functionality of bioprinted structures. By using algorithms and computer simulations, researchers can analyze the mechanical and biological properties of bioprinted structures, and make adjustments to improve their performance and functionality. This is particularly important in the field of tissue engineering, where the goal is to create functional structures that closely mimic the function and structure of native tissues.

One of the most promising applications of bioprinting and computational intelligence in tissue engineering and regenerative medicine is in the creation of functional organoids. Organoids are three-dimensional structures that closely mimic the function and structure of native organs, and can be used to study diseases, test new treatments, and even serve as a source of transplantable tissues. By using bioprinting technology to create these structures, researchers can create functional, three-dimensional models of diseased organs, which can be used to test new treatments and identify the most effective therapies for individual patients. Bioprinting and computational intelligence in tissue engineering and regenerative medicine is in the creation of functional, bioprinted prosthetics and implants. By using computational algorithms to analyze patient data, bioprinters can create custom-fitted implants that are tailored to the unique needs of each patient. This leads to improved outcomes and a higher quality of life for patients who require these devices.

Finally, bioprinting and computational intelligence are also being used to improve the outcomes of surgical procedures, such as tissue transplantation and regenerative medicine. By using bioprinted structures as a guide, surgeons can better understand the anatomy and function of the tissues they are working with, leading to safer and more effective procedures. In conclusion, bioprinting and computational intelligence are playing an increasingly important role in tissue engineering and regenerative medicine. By providing new and innovative solutions to some of the most challenging problems in healthcare, these technologies are helping to improve the quality of life for patients and advance the field of medicine. Whether it is through the creation of functional organoids, custom-fitted implants, or improved surgical procedures, bioprinting and computational intelligence are revolutionizing the way tissue engineering and regenerative medicine are approached.

## 1.7 Advancements in Bioprinting and Computational Intelligence Technologies

In recent years, there have been significant advancements in bioprinting and computational intelligence technologies. These advancements have opened up new and exciting possibilities for the field of tissue engineering and regenerative medicine, and have led to new and innovative solutions to some of the most challenging problems in healthcare.

One of the most notable advancements in bioprinting technology has been the development of new, more advanced bioprinters. These bioprinters are capable of creating three-dimensional structures with greater precision and accuracy, and can produce structures that are closer in function and structure to native tissues. This is particularly important in the field of regenerative medicine, where the goal is to create functional replacements for damaged or diseased tissues in the body.

The advancement in bioprinting technology is the development of new, more advanced biomaterials. Bioprinting requires the use of specialized materials that can support the growth and proliferation of cells, and that can provide the mechanical support required for functional tissues. Advances in biomaterials science have led to the development of new, more advanced materials that are more biocompatible and that support better cell growth and function.

In the field of computational intelligence, there have been significant advancements in the development of new, more advanced algorithms. These algorithms are used to analyze and optimize the design and functionality of bioprinted structures, and they are an essential component of the bioprinting process. Advances in algorithm design and implementation have led to the development of more sophisticated simulations and analyses, which provide a better understanding of the mechanical and biological properties of bioprinted structures.

Advancement in computational intelligence is the development of new, more advanced computer hardware. Computational intelligence requires large amounts of computing power, and advances in computer hardware have made it possible to run larger, more complex simulations and analyses. This has led to a better understanding of the design and functionality of bioprinted structures, and has paved the way for new and innovative solutions to some of the most challenging problems in tissue engineering and regenerative medicine.

In conclusion, there have been significant advancements in bioprinting and computational intelligence technologies in recent years. These advancements have opened up new and exciting possibilities for the field of tissue engineering and regenerative medicine, and have provided new and innovative solutions to some of the most challenging problems in healthcare. Whether it is through the development of more advanced bioprinters, new and improved biomaterials, or more sophisticated algorithms and computer hardware, bioprinting and computational intelligence are poised to revolutionize the way that of approach tissue engineering and regenerative medicine in the years to come.

## 1.8 The Ethical and Regulatory Implications of Bioprinting and Computational Intelligence

As bioprinting and computational intelligence technologies continue to advance and become more prevalent in the field of tissue engineering and regenerative medicine, it is important to consider the ethical and regulatory implications of these technologies. There are a number of important questions and concerns that must be addressed as these technologies continue to evolve, including issues related to patient safety, data privacy, and the equitable distribution of healthcare resources.

One of the most important ethical considerations in bioprinting is patient safety. Bioprinting involves the creation of functional tissues and organs using cells and biomaterials, and there is a risk that these tissues and organs may not function as intended. This is particularly concerning when it comes to implantable devices and transplants, as the failure of these devices can have serious and sometimes fatal consequences. As such, it is essential that bioprinting technologies be thoroughly tested and validated to ensure that they are safe and effective before they are used in clinical settings.

Ethical consideration in bioprinting is data privacy. Bioprinting requires the use of patient-specific information [4], such as medical images and genetic data, and there is a risk that this information may be misused or disclosed without the patient's consent. It is essential that bioprinting technologies be designed with strong data privacy and security controls, and that patients be fully informed about the use of their data and given control over how it is used.

In the field of computational intelligence, there are also important ethical considerations related to data privacy and the use of algorithms. Computational intelligence involves the use of large amounts of data and complex algorithms, and there is a risk that these algorithms may be biased or may not reflect the values and priorities of the individuals who are being treated. As such, it is essential that computational intelligence technologies be designed and validated to ensure that they are fair, transparent, and ethical.

In addition to these ethical considerations, there are also important regulatory implications of bioprinting and computational intelligence technologies. These technologies are still in their early stages of development, and there are currently no clear guidelines or regulations in place to govern their use. This can make it difficult for healthcare providers to know how to use these technologies in a safe and effective manner, and it can also limit the adoption and development of these technologies in the field of tissue engineering and regenerative medicine. In conclusion, as bioprinting and computational intelligence technologies continue to advance, it is important to consider the ethical and regulatory implications of these technologies. Issues related to patient safety, data privacy, and the equitable distribution of healthcare resources must be addressed, and clear guidelines and regulations must be put in place to ensure that these technologies are used in a safe and responsible manner. By addressing these issues and ensuring that these technologies are developed and used in an ethical and responsible manner, it is helpful to ensure that bioprinting and computational intelligence have a positive impact on the field of tissue engineering and regenerative medicine for years to come.

## 1.9 The Future of Bioprinting and Computational Intelligence: Opportunities and Challenges

Bioprinting and computational intelligence technologies have the potential to revolutionize the field of tissue engineering and regenerative medicine, and there is great excitement about the possibilities that these technologies offer. However, in order for these technologies to reach their full potential, there are also a number of important challenges that must be addressed. One of the greatest opportunities in the field of bioprinting is the potential to create functional tissues and organs that can be used for transplantation. This has the potential to revolutionize the way that treat a wide range of medical conditions, from heart disease to liver failure, by providing patients with functional, living tissues and organs that can be used to repair or replace damaged or diseased tissue.

The exciting opportunity in bioprinting is the potential to use these technologies for drug development and testing. Bioprinted tissues and organs can be used to create more accurate and predictive models of human physiology, and this can help researchers to develop new treatments and drugs more quickly and effectively. This has the potential to greatly accelerate the pace of medical research, and to help researchers to find new treatments for a wide range of conditions. In addition to these opportunities, there are also a number of important challenges that must be addressed in order for bioprinting and computational intelligence technologies to reach their full potential. One of the biggest challenges is the cost of these technologies. Bioprinting is a complex and time-consuming process, and the materials and equipment required are often expensive. As such, there is a need to find ways to reduce the cost of these technologies and make them more accessible to a wider range of patients.

Challenge in bioprinting is the need to improve the accuracy and reproducibility of these technologies. Bioprinting is still a relatively new field, and there is a need to improve the accuracy and consistency of these technologies to ensure that the tissues and organs that are created are of the highest quality. This requires further research and development to improve the materials, equipment, and processes used in bioprinting, and to ensure that these technologies are reliable and consistent.

In computational intelligence, there are also a number of important challenges that must be addressed. One of the biggest challenges is the need to ensure that these technologies are fair and ethical. As these technologies become more widely used, there is a risk that they may be biased or may not reflect the values and priorities of the individuals who are being treated. As such, there is a need to ensure that these technologies are designed and validated to ensure that they are fair, transparent, and ethical.

In conclusion, the future of bioprinting and computational intelligence is bright and full of exciting opportunities. However, in order for these technologies to reach their full potential, there are also a number of important challenges that must be addressed. From reducing the cost of these technologies to improving their accuracy and reliability, there is much work to be done to ensure that bioprinting and computational intelligence technologies have a positive impact on the field of tissue engineering and regenerative medicine for years to come.

## 1.10 Case Studies: Bioprinting and Computational Intelligence in Action

Bioprinting and computational intelligence technologies are being used in a growing number of innovative and exciting applications across a range of different fields. In this chapter, some of these case studies to see how these technologies are being used in real-world scenarios, and to gain a better understanding of the potential of bioprinting and computational intelligence is explored.

One of the most promising applications of bioprinting is in the field of tissue engineering and regenerative medicine. Bioprinting technologies have been used to create functional, living tissues and organs that can be used for transplantation, and this has the potential to revolutionize the way that this study treat a wide range of medical conditions. For example, researchers have used bioprinting technologies to create functional liver tissue, heart tissue, and blood vessels, which can be used to repair or replace damaged or diseased tissue in patients. The nest exciting application of bioprinting is in drug development and testing. Bioprinted tissues and organs can be used to create more accurate and predictive models of human physiology, which can help researchers to develop new treatments and drugs more quickly and effectively. For example, researchers have used bioprinted tissues to test the efficacy and safety of new drugs, which can help to reduce the time and cost of drug development and ensure that new treatments are safe and effective for patients. In computational intelligence, there are also a number of exciting case studies that demonstrate the potential of these technologies. For example, computational intelligence algorithms have been used to develop personalized treatment plans for patients with cancer, which can help to improve outcomes and ensure that patients receive the most effective treatments. The algorithms like CNN, SVM, DBNs can also be used to develop predictive models of patient outcomes, which can help clinicians to make more informed decisions about treatment and patient care.

In conclusion, the case studies explored in this chapter provide a glimpse into the exciting potential of bioprinting and computational intelligence technologies. From tissue engineering and regenerative medicine to drug development and testing, these technologies are being used in a wide range of innovative and exciting applications, and there is much to be gained from exploring their potential in greater detail.

### 1.10.1 Trends in Computational Intelligence and Bioprinting

The field of computational intelligence and bioprinting is rapidly evolving, and there are several trends that are shaping its development and growth. Here are some of the key trends that are worth paying attention to:

1. Increased Investment in Research and Development: There is growing recognition of the potential benefits of bioprinting and computational intelligence, and as a result, there is

increased investment in research and development in this area. This investment is driving rapid advances in these technologies, and is likely to lead to new applications and innovations in the years to come.

2. Development of Personalized Medicine: Bioprinting and computational intelligence are increasingly being used to develop personalized medicine, where treatments and therapies are tailored to the specific needs of each individual patient. This approach is expected to lead to improved patient outcomes and a more effective use of resources.
3. Advances in Tissue Engineering and Regenerative Medicine: Bioprinting is playing a growing role in tissue engineering and regenerative medicine, providing a means of creating functional, living tissues and organs that can be used for transplantation and drug development. Computational intelligence is also playing an important role in these fields, helping to optimize bioprinting technologies and make them more accurate and effective.
4. Growth of the Bioprinting Market: The bioprinting market is expected to grow rapidly in the coming years, driven by increased investment and the growing recognition of its potential benefits. This growth is expected to bring new opportunities for companies and researchers, as well as new innovations in the field.
5. Ethical and Regulatory Challenges: As bioprinting and computational intelligence continue to advance, there are also growing concerns about ethical and regulatory issues, such as the use of these technologies to create synthetic organisms or to manipulate human genes. These concerns are likely to become more pressing as these technologies continue to develop and gain wider use.
6. Integration with Artificial Intelligence and Machine Learning: Computational intelligence and bioprinting are increasingly being integrated with artificial intelligence and machine learning, creating new opportunities for automation and optimization of these technologies. This integration is expected to lead to faster, more accurate, and more effective bioprinting and computational intelligence technologies.

Overall, the trends in computational intelligence and bioprinting are positive, and suggest that these technologies are set to play an increasingly important role in shaping the future of medicine, healthcare, and many other industries. This study is helpful for a researcher, clinician, or simply someone who is interested in these technologies, it is important to stay up-to-date on the latest developments and trends in this exciting and rapidly evolving field.

### 1.10.2 Challenges in Computational Intelligence and Bioprinting

Despite the many benefits of computational intelligence and bioprinting, there are also significant challenges that must be addressed in order for these technologies to reach their full potential. Here are some of the key challenges facing the field also depicted in Figure 1.3:

1. Technical Challenges: Bioprinting is a complex and technically demanding process, requiring the precise control of multiple variables in order to produce functional, living tissues. Similarly, computational intelligence is a rapidly evolving field, requiring constant advances in algorithms and computing power in order to remain relevant. Overcoming these technical challenges is essential for the continued development and growth of these technologies.
2. Cost and Accessibility: Bioprinting technologies are still relatively expensive and inaccessible to many, making it difficult for researchers and clinicians to use these technologies on a wide scale. Furthermore, the cost of bioprinted tissues and organs is still high, making it difficult for patients to access these treatments. Addressing these issues will be critical for the widespread adoption of bioprinting and computational intelligence in the years to come.

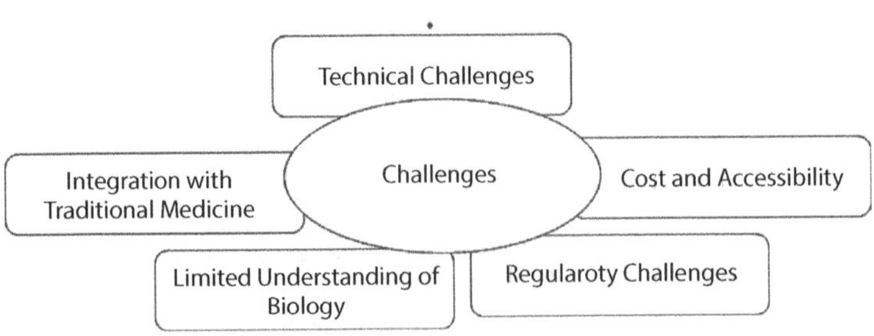


**Figure 1.3** Challenges of bioprinting.

3. Regulatory Challenges: Bioprinting and computational intelligence raise a number of important ethical and regulatory questions, such as the use of these technologies to create synthetic organisms or to manipulate human genes. Addressing these concerns will require careful consideration of the potential benefits and risks of these technologies, as well as the development of clear and effective regulations that balance these considerations.
4. Limited Understanding of Biology: Bioprinting and computational intelligence are highly interdisciplinary fields, requiring expertise in areas such as engineering, computer science, and biology. However, our understanding of biology is still limited, making it difficult to fully realize the potential of these technologies. Improving our understanding of biology and the biological processes that are involved in bioprinting and computational intelligence will be critical for the continued development of these technologies.
5. Integration with Traditional Medicine: [4] Bioprinting and computational intelligence represent a significant departure from traditional approaches to medicine, and there may be resistance to their adoption in some areas. Integrating these technologies with traditional medicine will require collaboration and communication between researchers, clinicians, and patients, as well as the development of effective training programs for healthcare professionals.

Despite these challenges, there is no doubt that computational intelligence and bioprinting represent a promising future for medicine, healthcare, and many other industries. By working to address these challenges, researchers and practitioners can help to ensure that these technologies reach their full potential, providing new treatments and therapies for patients and improving the quality of life for people around the world.

## 1.11 Conclusion

The promising intersection of bioprinting and computational intelligence bioprinting and computational intelligence are two technologies that are rapidly advancing and changing the way This study approaches many important medical and scientific problems. In this chapter, this study has explored the basics of these technologies and their role in shaping the future of medicine, tissue engineering, and drug development. This chapter have also discussed the ethical and regulatory implications of bioprinting and computational intelligence, and considered the opportunities and challenges that lie ahead.

The intersection of bioprinting and computational intelligence is particularly promising, as these technologies complement each other in many ways. Bioprinting provides a means of creating functional, living tissues and organs that can be used for transplantation and drug development, while computational intelligence provides the tools and algorithms that are needed to optimize these technologies and make them more accurate, predictive, and effective. The potential benefits of bioprinting and computational intelligence are many, and include improved patient outcomes, faster and more effective drug development, and the ability to create personalized treatment plans that are tailored to the needs of each individual patient. However, there are also many challenges that must be addressed, including ethical and regulatory issues, and the need for continued investment in research and development.

In conclusion, the intersection of bioprinting and computational intelligence is a field that is rich with promise, and one that is likely to play an increasingly important role in shaping the future of medicine and healthcare. This study deals with bioprinting with computational intelligence is beneficial for a researcher, clinician, or simply someone who is interested in these technologies, it is clear that bioprinting and computational intelligence are set to make a significant impact in the years ahead, and that there is much to be gained from exploring their potential in greater detail.

## References

1. Kumar, N., Jaganathan, S.K., Prabhu, K., 3D bioprinting: A review on its advancements and future prospects. *Biotechnol. Appl. Biochem.*, 2018.
2. Gungor-Ozkerim, P.S., Inci, I., Zhang, Y.S., Khademhosseini, A., Dokmeci, M.R., Bioinks for 3D bioprinting: An overview. *Biomater Sci.*, 1, 6, 5, 915–946, 2018.
3. Ng, W.L., Cherukuri, A., Yeong, W.Y., Bioprinting of human tissues: Current state-of-the-art, challenges, and opportunities. *Biotechnol. Adv.*, 2018.
4. Stanco, D., Urbán, P., Tirendi, S., Ciardelli, G., Barrero, J., 3D bioprinting for orthopaedic applications: Current advances, challenges and regulatory considerations. *Bioprinting*, 20, e00103, 2020
5. Liu, N., Ye, X., Yao, B., Zhao, M., Wu, P., Liu, G., ... Zhu, P., Advances in 3D bioprinting technology for cardiac tissue engineering and regeneration. *Bioactive Mater.*, 6, 5, 1388–1401, 2021.

2

# Design, Architecture, Implementation, and Evaluation of Bioprinting Technology for Tissue Engineering

**Vimala R. T. V.[1], Gangadevi E.[2]* and Lawanya Shri M.[3]**

[1]*Department of Bio-Technology,Bharathidasan University, Trichy, India*
[2]*Department of Computer Science, Loyola College, Chennai, India*
[3]*School of Information Technology and Engineering, Vellore, India*

***Abstract***
Recent advances in bioprinting have demonstrated promise in the fabrication of biostructures in the area of tissue engineering, especially orthopedic tissues and bones. Structural, mechanical as well as biological properties can be achieved with printed constructs, even when additive and traditional manufacturing processes are combined with the right biomaterials. There are still several improvements to be made, including improving printing resolution, speed, biomaterial compatibility, and expanding the variety of materials suitable for 3D printing. As part of this chapter, a comprehensive overview of current advancements in 3D bioprinting is provided, as well as guidelines for potential applications and material selection.

***Keywords:*** Bioprinting, 3D printing, tissue engineering, tissue regeneration

## 2.1 Introduction

In 3D bioprinting, biochemicals, biomaterials, and also living cells are precisely located with functional workings to fabricate tissues-like 3D structures [1]. Contrary to the conventional usage of 3D printing to produce cell-free scaffolds, the progress of 3D structures with mechanical and biological properties appropriate for the deposition of living cells and the rebuilding of tissue and organ function using 3D bioprinting necessitates

**Corresponding author*: smgangadevi@gmail.com

E. Gangadevi, M. Lawanya Shri, Rajesh Kumar Dhanaraj, and Balamurugan Balusamy (eds.) Computational Intelligence in Bioprinting, (21–38) 

the use of numerous technical approaches, including mini-tissue building blocks, and biomimicry [2].

3D bioprinting has several advantages over traditional 3D printing, including scalability, cost, high-resolution cell deposition, and accurate cell distribution. However, there are numerous challenges to the development and subsequent applications of 3D bioprinting in a variety of industries. For example, present printing techniques must be enhanced for greater scalability; faster printing speeds and even advanced printing resolution are required to make specific biological functions without sacrificing automatic properties.

In current years, a variety of outstanding review articles on the advanced development of 3D bioprinting have been published. Nearly all facets of 3D bioprinting, including principles, imaging, techniques, and materials choice, were covered [2]. Aside from providing examples of how conventional non-biological 3D printing methods have been improved for better biocompatibility and biomaterial fabrication, articles on advances in 3D printing skills and materials are also available [3].

This chapter discusses significant advancements in 3D bioprinting-based tissue engineering. We will first discuss the primary 3D printing techniques, along with their benefits and drawbacks. The requirements for biomaterials that can be printed in three dimensions are then discussed. Additionally, we evaluate the requirements for bioink design for extrusion-based 3D bioprinting methods that are used to develop multiple constructs. Evolving trends and future directions in bioinks and bioprinting are also

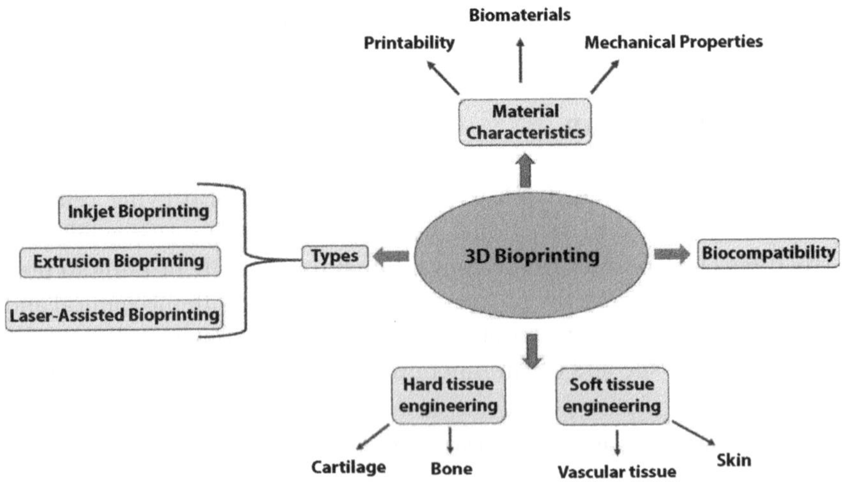


**Figure 2.1** Schematic depiction of application of 3D Bioprinting in Biomedical field.

emphasized. Understanding mechanical properties in the bioprinting process is important to all of us, as is knowing how processing parameters affect cellular viability. Finally, a list of promising new bioprinting research directions is formed. The potential of 3D bioprinting in medicine has also been studied (Figure 2.1).

## 2.2 3D Bioprinting

Traditionally, solid scaffolds are developed using thermoplastics or resins as biomaterials on the 3D printer [4]. This system either polymerizes liquid resin, heats filament as it passes through a nozzle, or sinters powdered material. The traditional tissue engineering principle is to seed isolated cells on pre-formed solid scaffolds, place the construct inside a bioreactor, and then implant it to patients.

3D printing has some limitations when it comes to hollow organs despite its capacity for producing hollow organs. Complex 3D organ reconstruction is difficult due to nonuniform cell seeding, the placement of many types of cells in defined positions, and the inability to place cells in defined positions. If cell seeding is not similar across the scaffold, imperfect healing may occur. High temperatures and toxic solvents can also reduce cell survival. Furthermore, a longer time duration for cell expansion may delay the treatment. As a result, the techniques' qualities should be investigated in terms of critical criteria like printing resolution, and the substance used to generate the 3D structures. Additionally, it is very challenging to develop a functional organ with clinical dimensions. The challenges include integrating the vascular network structure, and functionality as well as combining various cell types to prepare complex organ geometry.

3D bioprinting developed as a result of improvements made to the current 3D technology for tissue engineering. Recently, the use of 3D bioprinting, which quickly builds complex 3D objects from the top down while directly depositing living cells, has gained popularity. The basis of 3D printing is to assemble biomaterials layer-by-layer to create necessary geometry, which is developed from additive manufacturing. It has become increasingly popular as a method of directly putting living cells into 3D complex objects using a top-down approach. It works by layering biomaterials to prepare the geometry required, which evolved from additive manufacturing. There are many potential applications for 3D printing techniques in tissue engineering and regenerative drug.

Nowadays, bioprinting enables the incorporation of cells with biomaterials such as ink by printing tissue constructs in ambient conditions using hydrogel material. By arranging cells, macro molecules, extracellular components, and biomaterials in a predetermined pattern in three or two dimensions to mimic functional tissue, the current 3D Bioprinting technique incapacitates the drawbacks of 3D printing. `For the purpose of developing tissues, organ-on-a-chip models, or tissue constructs, bioprinting is the simultaneous printing of proteins, DNA, drug particles, biomaterials, and progress factors in a specified layer-by-layer order. Among the biomaterials that can be used for 3D bioprinting are hydrogels, gelatin from the extracellular matrix, proteins, and polymers. The one-of-a-kind benefit of designing scaffold with 3D bioprinting is that it eliminates the issue of a smaller number of donors availability. The benefit of designing scaffold to 3D bioprinting addresses the problem of limited donor availability by printing tissue constructs with cell-infused bioink. The advantages of bioprinting technology include *in situ* cell printing, accurate cell positioning, and the capabilities to print tissues with heavy cell densities and vascularization.

The main objective remains to identify biological materials that are both suitable for 3D printing and satisfy the mechanical and functional necessities for tissue constructs. This section will first discuss the ideal features of materials, followed by an examination of some biomaterials as examples.

## 2.3 Material Characteristics

### 2.3.1 Printability

One of the important characteristics of a substance that qualifies it for 3D bioprinting is its capability to be fully used by the printer, precisely how the material can be placed with sufficient controllability. However, printability varies depending on the printing method used. For example, Inkjet printing is limited by material viscosity, extrusion-based printing can print substances with extremely high viscosity, although this requires a specialized inter-layer crosslinking mechanism. However, some procedures require severe heating of the material for cell deposition, it is necessary for the printing element to shield the cells as this high temperature. Materials with low heat conductivity or those that may cushion cells through the deposition process are more likely to increase cell survival and biological function after printing [5]. Surface tension amid the printing media and the receiving substrate is additional factor that influences cell adherence and development [6].

## 2.4 Mechanical Properties

The material must possess sufficient mechanical and structural qualities to support a three-dimensional structure after solidifying. Cells need to have a robust structure to adhere, grow, and differentiate in a healthy environment [7]. Moreover, interactions between cells as well as the printing medium have been exposed to have a significant impact on cell adhesion [8]. Additionally, mechanical requirements for materials vary based on the kind of tissue engineering, which includes hard implanted bone as well as soft tissues like skin and cartilage; mechanical qualities play an important part in soft tissue activities.

## 2.5 Biomaterials

Both synthetic and natural polymers (fibronectin, and alginate) are employed in 3D bioprinting [9]. Natural polymers will have excellent cell interactions. As a result, extracellular decellularization appears to be a promising method for creating intact decellularized matrices for use in bioprinting processes. In previous study, bioinks will successfully printed from three different tissues using dECM. The properties and biological functions of the bioinks' dECM compositions differ from tissue to tissue and may be related.

On the other hand, the creation of materials through chemical synthesis, results in materials with distinctive mechanical and chemical properties that can be tailored for a range of bioprinting applications. For example, pluronics are triblock copolymers with a hydrophilic polyethylene glycol A block (PEG) and a hydrophobic PEG B block. Pluronics is suitable for 3D printing owing to its low gelation temperature and high printability [10]. PEG is also extensively used in many 3D bioprinting compositions, either to develop hydrogels or cross linkable polymers after functionalization with a polymerization agent. However, they are not developed with the goal of maintaining cell viability over the long term and lack the bioactivity of synthetic polymers [11].

## 2.6 Design, Architecture of 3D Bioprinting

The fascinating new field of bioprinting appears to have applications in both medicine and science. There has been significant progress in the design of

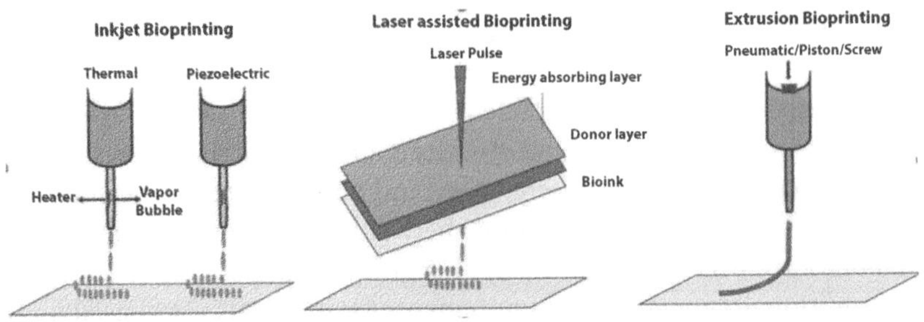


**Figure 2.2** Common bioprinting techniques.

complex biomimetic constructs with cellular functions. Therefore, designing of complex material is categorized as show below (Figure 2.2).

### 2.6.1 Inkjet Bioprinting

The inkjet bioprinting, is developed from the procedures of standard inkjet printing [12]. Inkjet bioprinting, unlike standard 2D inkjet printing, employs bioink in the form of a hydrogel pre-polymer solution containing encapsulated cells. In which, droplets of bioink are ejected onto a platform's substrate. Layer by layer, the process is continued indefinitely to find patterns, and solidify to develop the 3D object. The head of the printer produces controllable-size drops by applying thermal or piezoelectric force [13]. A voltage is set to the crystal substance to split the bioink into droplets after the nozzle at predetermined time throughout the printing process [14]. The resolution and mechanical characteristics of inkjet-printed structures are well known to be influenced by the cell density in the nozzle, size, and bioink. Inkjet bioprinting has several advantages, including simplicity of use, low cost, and fast printing [15]. Further, the stability of biological molecules like DNA, as well as the viability of mammalian cells, are unaffected by this temperature variation [16].

Recently, printed constructs are programmed with cell function using therapeutic-loaded bioinks. For example, a poly (ethylene glycol)-dithiothreitol bioink and 2D nanosilicates with protein therapies revealed a shear-thinning rheological characteristic and increased printing quality. 3D printed structures can sequester and release protein therapeutics due to the direction-specific interactions of 2D nanosilicates (Figure 2.3a). This method offers the potential for engineering complex 3D tissue structures for regenerative medicine [17].

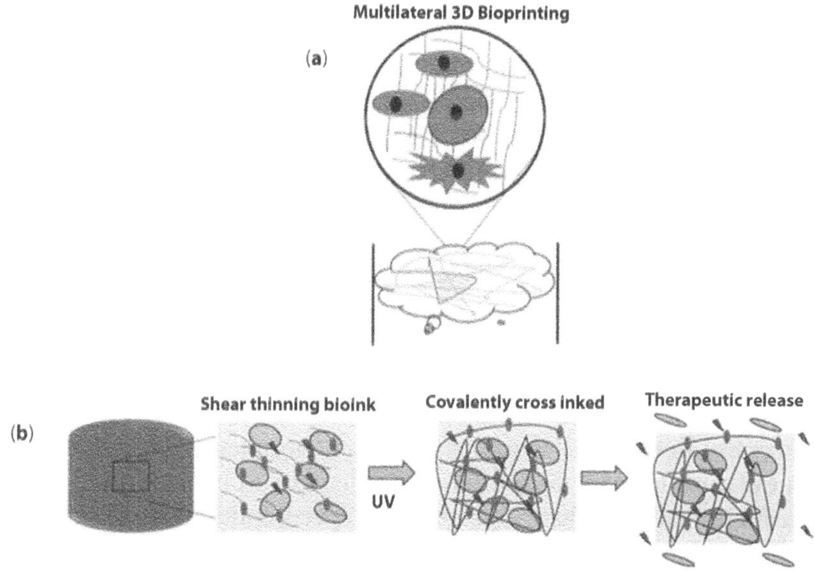


**Figure 2.3** (a) Multilateral 3D bioprinting and (b) therapeutics.

## 2.6.2 Laser-Assisted Bioprinting (LAB)

Direct-write and laser-induced transfer advancement is referred to as LAB. In LAB, on the bottom of the donor's "ribbon" is a layer of bioink (hydrogel or cells), and on top is an energy-absorbing layer such as gold or titanium. During printing, the laser source produces laser pulses that stimulate the energy-absorbing layer. Due to energy absorption, a part of the donor layer vaporizes during this stage, launching bioink droplets onto the substrate with a high pressure bubble. The properties of LAB-printed material are influenced by surface tension, thickness, and density as well as laser wavelength, intensity, and pulse time [18].

As opposed to other bioprinting methods, LAB uses no nozzles and does not require any contact. LAB printing causes no mechanical pressure to the cells, ensuing in high cell viability. LAB can print a wide range of biological materials, such as high viscosity biological materials, mammalian cells, and high-density cells [19, 20]. LAB offers a important advantage over other bioprinting technologies since nozzle obstruction may be avoided. However, preparing the "ribbon" set up for every cell or hydrogel type takes time, especially when many cell types are made [21].

### 2.6.3 Extrusion Bioprinting

Extrusion printing is one of the most inexpensive techniques for rapid prototyping [22]. Extrusion bioprinting is a more advanced form of inkjet bioprinting as it permits the deposit of extremely viscous materials and dense cells to construct 3D structures. A micro-nozzle is used to extrude a continuous line of bioink under the influence of a pneumatic piston instead of using liquid droplets. The extruded material act as a supporting structural material after solidifying on the substrate, the platform is then reduced horizontally, and the next layer of bioink has been mixed until the entire 3D construct is completed.

In which, mechanical dispensing printers allow direct control over the substance flow than pneumatically driven printers, resulting in higher spatial control due to a delay in the gas volume in pneumatic systems [23]. Some of the important factors to reflect when extrusion bioprinting to produce high-quality results are the extrusion speed, and density of the bioink, the liquid phase of the bioink and other properties.

Extrusion bioprinting offers a number of significant benefits, one of which is the capability to add a high density of cells, allowing a wider range of substance selection with many kinds of cell densities and viscosities. In extrusion bioprinting, high-viscous natured material will provide structural care with printed components, whereas low-viscous natured material can be used to develop an environment opted for preserving cell function and their viability [24]. One of the more sophisticated features described below is multilateral 3D bioprinting, a recent development.

## 2.7 3D Bioprinting Tissue Models

3D Biofabricated tissue structures are becoming more and more common as a platform for studying different diseases and evaluating the effectiveness of cutting-edge therapeutic strategies, potentially reducing or replacing the need for animal testing [25]. Traditional fabrication methods, however, frequently result in overly simplistic constructs and cell microenvironments [26]. 3D bioprinting makes it likely to create complex, biomimetic tissue representations that can contribute to the optimization of treatments [27]. For instance, it is thought that the microenvironment is essential for comprehending and regulating the growth and metastasis of tumors. 3D bioprinted tumor models are very useful for preclinical research even though they allow for more precise modeling of the tumor environment (Figure 2.3b).

## 2.8 3D Multimaterial Bioprinting- Development of Complex Architectures

The intricate compositions of native tissue structures cannot be replicated by current printing technologies. The development of several additive manufacturing techniques that can print relatively complex geometries has led to the emergence of multi material bioprinting [28, 29]. In multiple material, bioinks are continuously deposited in a coded manner, allowing seamless transitions between materials in tissue constructs. This enables printing structures to be produced that closely resemble the designs and composition of native tissues [30]. In which, multiple bioinks can be deposited in spatially defined 3D architectures that are calibrated by a motor-driven stage movement. In spite of this, the development of additive manufacturing systems that can print complex architectures precisely remains a significant challenge [31].

## 2.9 Implementation and Evaluation

3D bioprinting is becoming a enormous industry due to its versatility and potential applications. Bioprinting is primarily used for tissue regeneration, such as liver, skin, nerves, and heart valves [32]. These are described more below.

Hard tissue engineering

Hard tissue (bone and cartilage) regeneration may be possible with 3D bioprinting. Hard tissue engineering necessitates the use of a mineralized extracellular matrix to mimic the tissue environment [33]. Bone is made up of osteoblasts and the cartilage is composed of chondrocytes, are two distinct cell types in hard tissues. Hematopoiesis and metabolism are the biological functions of hard tissues. Hard tissues need mechanical support, and the biomaterial used should offer strength to the resulting product.

## 2.10 Bone

Bone is a kind of tissue with cemented ECM. Osteoblasts, osteocytes, and osteoclasts are the three major cell types of Bone. Biomaterials used must be printable, biocompatible, and promote uniform cell dispersion. Therefore, Biomaterial selection and functional vasculature are the most difficulties in bone bioprinting. All these problems can be solved by using the proper bioprinting technique [34]. In other studies, Gelatin-methacryloyl hydrogel was used as the bioink with low methacryloyl substitution for direct-writing extrusion-based bioprinting on bone [35].

Natural bone's main building block, hydroxyapatite, contains osteo conductivity, which assists in developing a microenvironment that is like bone. Hydroxyapatite-containing hydrogels increased cell proliferation and osteogenic differentiation [36]. In addition, the researcher was combined alginate/hydroxyapatite with Atsttrin's antagonist effect to repair complex calvarial bone defects and improve TNF/TNFR signals. Moreover, a new hydrogel was developed from mouse calvaria 3T3-E1 (MC3T3)TNFR signals using polyvinyl alcohol, alginate, and hydroxyapatite. The porous scaffold that was created accelerates and uniformly heals bone defects [37, 38].

Hyaluronic acid (HA) maintains the extracellular matrix mechanically. In which, a 3D printed scaffold was developed through the combination of photoinitiators and methacrylate groups with HA [39]. The resulting methacrylated HA gels are mechanically stiff, with poor cytocompatibility, good biocompatibility, and resistance to degradation. They also aid in the survival of primordial cells. As a consequence, the prerequisites for bone tissue engineering are achieved [40].

Titanium is a biocompatible material that may be utilized as biological implants. However, when it comes into contact with tissues, the load is transmitted into the implants, causing them to loosen over time without developing antigenicity. Therefore, titanium implant osseointegration is required for a functional connection. In recent research, titanium implant osseointegration was enhanced by developing a gelatin metha cryloyl (GelMA) scaffold with the help of extrusion 3D bioprinter. This scaffold can be grafted onto a titanium implant or directly printed on it [41].

Bioactive glasses may stimulate angiogenesis, bone cell mineralization, and oesteogenesis. Wang and colleagues revealed into how bioglass affected the biomineralization of bone-related SaOS-2 cells. Furthermore, bioactive glasses were supplemented with alginate/gelatin as hydrogel. Bioglass enhances the proliferative and mineralization abilities of SaOS-2 cells [42]. In another study, bioactive glass and hydroxyapatite were printed using hMSC suspensions postponed in poly (ethylene glycol) di-metha crylate to examine their osteogenesis characteristics [43].

A two-syringe extrusion bioprinting method (bioactive glass), one with human adipose stem cells in Matrigel and the other with PCL polymer, was developed. A layer of bioink is printed first, and then a syringe with bioactive glass is allowed to print. Three-dimensional bioprinted polycaprolactone/bioactive borate glass scaffolds provide improved angiogenesis for bone tissue engineering [44]. Gao *et al.* created hMSCS-incorporated PEGGelMA hydrogel scaffold using an inkjet bioprinter with simultaneous photopolymerization capability. This scaffold exhibits increased mechanical power, which is necessary to the engineering of bone and cartilage tissue [43].

## 2.11 Cartilage

A connective tissue called cartilage lacks a nervous and circulatory system. It contains the specialized cells known as chondrocytes. The lack of blood vessels causes nutrition to diffuse through the matrix, which leads to a slow capacity for self-healing. There are three kinds of cartilage: fibrous, hyaline, and elastics [45].

Chondrogenic cell-based bioink is favored to produce synthetic cartilage. Therefore, synthetic cartilage for example, materials containing methacrylate and alginate are combined with suitable hydrogel and compatible cells for use as bioink for the purpose of promoting cartilage-based tissue engineering and chondrogenesis. Further, several research investigations have been carried out to support the incorporation of the chondrocyte phenotype into alginate-blended bioink. For example, a multi-head deposition device was used to synthesize a hybrid polycaprolactone scaffold with chondrocyte cell-encapsulated alginate hydrogel [46]. Nanofibrillated cellulose and alginate were used to bioprint this hydrogel, which was then filled with chondrocyte cells to form cartilage [13].

*In vivo* human cartilage was developed using an extrusion printer by joining nanocellulose and alginate [47]. In recent studies, polyethylene glycol and methacrylated poly(N-(2-hydroxypropyl) metha crylamide mono/dilactate) triblock copolymers was developed as a hydrogel, This blend of polymers encourages the growth of cartilage while also having strong mechanical properties [48]. Chondrocyte-laden bioprinter Gelatin-metha cryloyl (GelMA) was mixed with gellan gum to determine whether it could be used as bioink for cartilage fabrication. Furthermore, this hydrogel was discovered to promote to reduce viscosity, resulting in the hydrogel's high printability [49].

## 2.12 Soft Tissue Engineering

Treatment of defects with a high volume to surface ratio is the primary goal of soft tissue engineering. Surgery and autologous tissue transfer, which are the two most common therapeutic modalities, both cause morbidity at the donor site at the injury site.

## 2.13 Vascular Tissue

The tunica external layer, tunica middle layer, and tunica internal layer are the three main layers that make up the vascular tissue. The innermost layer is made up of endothelial cells, which offer anti-thrombogenic and anti-infection properties. The following layer is the basement layer, which is composed of collagen type IV and laminin. Fibroblasts make up the tunica adventitia, the skin's outermost layer. Each layer cooperates with the others to carry out its specific tasks. Based on the size of the vessels, the vascular system is divided into macrovasculature and microvasculature. Arterioles, capillaries, and venules make up the microvasculature, whereas arteries and veins make up the microvasculature [50].

Vascularity is needed for the proper function of any tissue construct such as tissues and organs. Vessel-like structure is necessary to improve functional tissue matching for tissue regeneration. Despite being biocompatible and encouraging cell growth, Extracellular matrix-based biomaterials are not strong enough to support the weight of tubes despite being biocompatible. Therefore, they were combined with a suitable biomaterial that had mechanical stability to develop highly organized and intact vessels. The interconnected networks within tissue constructs facilitate the transportation of oxygen, nutrients, and have numerous applications [51].

Moreover, traditional microfabrication techniques are incapable of producing perusable hollow vascular tubes, resulting in the creation of necrotic regions in bulk tissues. The vascular network by deposition of patterned cells and biomaterials was developed using extrusion system. For example, cell responsive bioink comprising gelatin methacryloyl (GelMA), and 4-arm poly(ethylene glycol)-tetra-acrylate was developed recently [52]. Similarly, cell-laden ink, fugitive ink, silicone ink as bioink was developed within 3D perfusion chips to print thick vascularized tissue [53]. The network structure follows different mechanisms for the storage capacity [54]. Pati *et al.* used bone marrow-derived mesenchymal stem cells in sodium alginate on cross-stripped PCL scaffold layer to bioprint a small diameter vessel [55].

## 2.14 Skin

In the case of extensive burn injury, conventional treatment techniques such as autografts and allografts are difficult to use. This treatment option has a number of drawbacks, including limited size, a lengthy preparation time, and contracture. As a result, the bioprinting technique can be used to print artificial skin by incorporating multiple kinds of skin cells.

Inkjet bioprinting has recently been planned as a method for directly printing keratinocytes onto the wound model for *in situ* skin repair. Skin produced using these methods is utterly identical to typical murine skin [56]. In other studies, skin in a mouse wound was regenerated using a printed collagen-fibrin gel comprising AFS cells from amniotic fluid. The developed construct promotes the growth of blood vessel and neovascularization and the presence of AFS cells modifies the inflammatory response [57].

## 2.15 Biocompatibility and Control of Degradation and Byproducts

The term "biocompatibility" refers to a substance to behave in a particular situation while eliciting an appropriate host response [58]. It has become more about passively allowing or active producing positive effects in the host rather than the implantation substance to coexist with the host through the evolution of biocompatibility [59]. Further, the material should allow the cells to replace it with their own ECM proteins at a rate that matches the material's rate of degradation after implanting into the host. Therefore, it is confirmed that the biocompatibility of a material is also determined by the byproducts produced during the degradation process [60].

## 2.16 Conclusion

In the context of bioprinting, additive manufacturing has enormous potential for tissue and organ regeneration. Even though 3D bioprinting produces tissue from scratch, it presents a unique opportunity for reducing donor shortage and reducing the possibility of graft rejection. The technology of 3D bioprinting has advanced to fulfill the needs of the tissue engineering industry. This is since it has introduced numerous approaches to tissue fabrication, such as inkjet printing, extrusion bioprinting, laser-assisted bioprinting, and so on. Comparatively speaking to manual tissue culture techniques, 3D bioprinting techniques enable advanced tissue printing with finely tuned cell patterning and improved spatial control.

Despite all the developments, the printed construct's biocompatibility and body integration remain significant obstacles. Cell viability must be maintained in the bioink preparation and then printed in geometries to maintain the quality of the printed construct. This necessitates the standardization of printing methods as well as meticulous quality control.

## References


1. Mandrycky, C., Wang, Z., Kim, K., Kim, D.-H., 3D bioprinting for engineering complex tissues. *Biotechnol. Adv.*, 34, 4, 422–434, 2016.
2. Murphy, S.V. and Atala, A., 3D bioprinting of tissues and organs. *Nat. Biotechnol.*, 32, 8, 773–785, 2014.
3. Chia, H.N. and Wu, B.M., Recent advances in 3D printing of biomaterials. *J. Biol. Eng.*, 9, 1, 1–14, 2015.
4. Gao, G., Schilling, A.F., Hubbell, K., Yonezawa, T., Truong, D., Hong, Y., Dai, G., Cui, X., Improved properties of bone and cartilage tissue from 3D inkjet-bioprinted human mesenchymal stem cells by simultaneous deposition and photocrosslinking in PEG-GelMA. *Biotechnol. Lett.*, 37, 2349–2355, 2015.
5. Talbot, E.L., Berson, A., Brown, P.S., Bain, C.D., Evaporation of picoliter droplets on surfaces with a range of wettabilities and thermal conductivities. *Phys. Rev. E Stat. Nonlinear Soft Matter Phys.*, 85, 061604, 2012.
6. Duymaz, B.T., Erdiler, F.B., Alan, T., Aydogdu, M.O., Inan, A.T., Ekren, N., Uzun, M., Sahin, Y.M., Bulus, E., Oktar, F.N., 3D bio-printing of levan/polycaprolactone/gelatin blends for bone tissue engineering: Characterization of the cellular behavior. *Eur. Polym. J.*, 119, 426–437, 2019.
7. Murphy, S.V., Skardal, A., Atala, A., Evaluation of hydrogels for bio-printing applications. *J. Biomed. Mater. Res. A*, 101, 1, 272–284, 2013.
8. Benson, K., Galla, H.J., Kehr, N.S., Ce ll adhesion behavior in 3D hydrogel scaffolds functionalized with d-or l-aminoacids. *Macromol. Biosci.*, 14, 6, 793–798, 2014.
9. Lee, K.W., Kim, D.-H., Lee, J.H., Youn, Y.-N., The effect of pulsatile flow on bMSC-derived endothelial-like cells in a small-sized artificial vessel made by 3-dimensional bioprinting. *Stem Cells Int.*, 2018, 1–11, 2018.
10. Donderwinkel, I., Van Hest, J.C., Cameron, N.R., Bio-inks for 3D bioprinting: Recent advances and future prospects. *Polym. Chem.*, 8, 31, 4451–4471, 2017.
11. Suntornnond, R., An, J., Chua, C.K., Bioprinting of thermoresponsive hydrogels for next generation tissue engineering: A review. *Macromol. Mater. Eng.*, 302, 1, 1600266, 2017.
12. Singh, M., Haverinen, H.M., Dhagat, P., Jabbour, G.E., Inkjet printing—Process and its applications. *Adv. Mater.*, 22, 6, 673–685, 2010.
13. Markstedt, K., Mantas, A., Tournier, I., Martínez Ávila, H., Hagg, D., Gatenholm, P., 3D bioprinting human chondrocytes with

nanocellulose–alginate bioink for cartilage tissue engineering applications. *Biomacromolecules*, 16, 5, 1489–1496, 2015.

14. Tekin, E., Smith, P.J., Schubert, U.S., Inkjet printing as a deposition and patterning tool for polymers and inorganic particles. *Soft Matter*, 4, 4, 703–713, 2008.
15. Orloff, N.D., Truong, C., Cira, N., Koo, S., Hamilton, A., Choi, S., Wu, V., Riedel-Kruse, I.H., Integrated bioprinting and imaging for scalable, networkable desktop experimentation. *RSC Adv.*, 4, 65, 34721–34728, 2014.
16. Xu, T., Gregory, C.A., Molnar, P., Cui, X., Jalota, S., Bhaduri, S.B., Boland, T., Viability and electrophysiology of neural cell structures generated by the inkjet printing method. *Biomaterials*, 27, 19, 3580–3588, 2006.
17. Cross, L.M., Carrow, J.K., Ding, X., Singh, K.A., Gaharwar, A.K., Sustained and prolonged delivery of protein therapeutics from two-dimensional nanosilicates. *ACS Appl. Mater. Interfaces*, 11, 7, 6741–6750, 2019.
18. Koch, L., Kuhn, S., Sorg, H., Gruene, M., Schlie, S., Gaebel, R., Polchow, B., Reimers, K., Stoelting, S., Ma, N., Laser printing of skin cells and human stem cells. *Tissue Eng. Part C Methods*, 16, 5, 847–854, 2010.
19. Gruene, M., Deiwick, A., Koch, L., Schlie, S., Unger, C., Hofmann, N., Bernemann, I., Glasmacher, B., Chichkov, B., Laser printing of stem cells for biofabrication of scaffold-free autologous grafts. *Tissue Eng. Part C Methods*, 17, 1, 79–87, 2011.
20. Koch, L., Deiwick, A., Schlie, S., Michael, S., Gruene, M., Coger, V., Zychlinski, D., Schambach, A., Reimers, K., Vogt, P.M., Skin tissue generation by laser cell printing. *Biotechnol. Bioeng.*, 109, 7, 1855–1863, 2012.
21. Müller, M., Öztürk, E., Arlov, Ø., Gatenholm, P., Zenobi-Wong, M., Alginate sulfate–nanocellulose bioinks for cartilage bioprinting applications. *Ann. Biomed. Eng.*, 45, 210–223, 2017.
22. Sears, N.A., Seshadri, D.R., Dhavalikar, P.S., Cosgriff-Hernandez, E., A review of three-dimensional printing in tissue engineering. *Tissue Eng. Part B Rev.*, 22, 4, 298–310, 2016.
23. Chang, C.C., Boland, E.D., Williams, S.K., Hoying, J.B., Direct-write bioprinting three-dimensional biohybrid systems for future regenerative therapies. *J. Biomed. Mater. Res. Part B Appl. Biomater.*, 98, 1, 160–170, 2011.
24. Emmermacher, J., Spura, D., Cziommer, J., Kilian, D., Wollborn, T., Fritsching, U., Steingroewer, J., Walther, T., Gelinsky, M., Lode, A., Engineering considerations on extrusion-based bioprinting: Interactions of material behavior, mechanical forces and cells in the printing needle. *Biofabrication*, 12, 2, 025022, 2020.
25. Gold, K., Gaharwar, A.K., Jain, A., Emerging trends in multiscale modeling of vascular pathophysiology: Organon-a-chip and 3D printing. *Biomaterials*, 196, 2, 2019.
26. Jang, J., Yi, H.-G., Cho, D.-W., 3D printed tissue models: Present and future. *ACS Biomater. Sci. Eng.*, 2, 1722, 2016.

27. Nguyen, D.G., Funk, J., Robbins, J.B., Crogan-Grundy, C., Presnell, S.C., Singer, T., Roth, A.B., Bioprinted 3D primary liver tissues allow assessment of organ-level response to clinical drug induced toxicity *in vitro*. *PloS One*, 11, 7, e0158674, 2016.
28. Joung, D., Truong, V., Neitzke, C.C., Guo, S.Z., Walsh, P.J., Monat, J.R., Meng, F., Park, S.H., Dutton, J.R., Parr, A.M., 3D printed stem-cell derived neural progenitors generate spinal cord scaffolds. *Adv. Funct. Mater.*, 28, 39, 1801850, 2018.
29. Liu, W., Zhang, Y.S., Heinrich, M.A., De Ferrari, F., Jang, H.L., Bakht, S.M., Alvarez, M.M., Yang, J., Li, Y.C., Trujillo-de Santiago, G., Rapid continuous multimaterial extrusion bioprinting. *Advanced Mater.*, 29, 3, 1604630, 2017.
30. Idaszek, J., Costantini, M., Karlsen, T.A., Jaroszewicz, J., Colosi, C., Testa, S., Fornetti, E., Bernardini, S., Seta, M., Kasarełło, K., 3D bioprinting of hydrogel constructs with cell and material gradients for the regeneration of full-thickness chondral defect using a microfluidic printing head. *Biofabrication*, 11, 4, 044101, 2019.
31. Han, X., Bibb, R., Harris, R., Engineering design of artificial vascular junctions for 3D printing. *Biofabrication*, 8, 2, 025018, 2016.
32. Derakhshanfar, S., Mbeleck, R., Xu, K., Zhang, X., Zhong, W., Xing, M., 3D bioprinting for biomedical devices and tissue engineering: A review of recent trends and advances. *Bioact. Mater.*, 3, 2, 144–156, 2018.
33. Sun, T., Yao, S., Liu, M., Yang, Y., Ji, Y., Cui, W., Qu, Y., Guo, X., Composite scaffolds of mineralized natural extracellular matrix on true bone ceramic induce bone regeneration through Smad1/5/8 and ERK1/2 pathways. *Tissue Eng. Part A*, 24, 5-6, 502–515, 2018.
34. Keriquel, V., Oliveira, H., Rémy, M., Ziane, S., Delmond, S., Rousseau, B., Rey, S., Catros, S., Amédée, J., Guillemot, F., *In situ* printing of mesenchymal stromal cells, by laser-assisted bioprinting, for *in vivo* bone regeneration applications. *Sci. Rep.*, 7, 1, 1–10, 2017.
35. Byambaa, B., Annabi, N., Yue, K., Trujillo-de Santiago, G., Alvarez, M.M., Jia, W., Kazemzadeh-Narbat, M., Shin, S.R., Tamayol, A., Khademhosseini, A., Bioprinted osteogenic and vasculogenic patterns for engineering 3D bone tissue. *Adv. Healthc. Mater.*, 6, 16, 1700015, 2017.
36. Demirtaş, T.T., Irmak, G., Gümüşderelioğlu, M., A bioprintable form of chitosan hydrogel for bone tissue engineering. *Biofabrication*, 9, 3, 035003, 2017.
37. Bendtsen, S.T., Quinnell, S.P., Wei, M., Development of a novel alginate-polyvinyl alcohol-hydroxyapatite hydrogel for 3D bioprinting bone tissue engineered scaffolds. *J. Biomed. Mater. Res. A*, 105, 5, 1457–1468, 2017.
38. Wang, X., Tolba, E., Schröder, H.C., Neufurth, M., Feng, Q., Diehl-Seifert, B., Müller, W.E., Effect of bioglass on growth and biomineralization of SaOS-2 cells in hydrogel after 3D cell bioprinting. *PloS One*, 9, 11, e112497, 2014.

39. Poldervaart, M.T., Goversen, B., De Ruijter, M., Abbadessa, A., Melchels, F.P., Öner, F.C., Dhert, W.J., Vermonden, T., Alblas, J., 3D bioprinting of methacrylated hyaluronic acid (MeHA) hydrogel with intrinsic osteogenicity. *PloS One*, 12, 6, e0177628, 2017.
40. Amini, A. R., Laurencin, C. T., Nukavarapu, S. P., Bone tissue engineering: Recent advances and challenges, in: *Critical Reviews in Biomedical Engineering*, vol. 40, issue 5, pp. 363–408, Begell House, 2012, https://doi.org/10.1615/critrevbiomedeng.v40.i5.10
41. McBeth, C., Lauer, J., Ottersbach, M., Campbell, J., Sharon, A., Sauer-Budge, A.F., 3D bioprinting of GelMA scaffolds triggers mineral deposition by primary human osteoblasts. *Biofabrication*, 9, 1, 015009, 2017.
42. Wang, Y., Application of 3D bioprinting in cartilage tissue. *AIP Conference Proceedings*, AIP Publishing LLC, p. 020047, 2019.
43. Gao, G., Zhang, X.F., Hubbell, K., Cui, X., NR2F2 regulates chondrogenesis of human mesenchymal stem cells in bioprinted cartilage. *Biotechnol. Bioeng.*, 114, 1, 208–216, 2017.
44. Mouser, V.H., Melchels, F.P., Visser, J., Dhert, W.J., Gawlitta, D., Malda, J., Yield stress determines bioprintability of hydrogels based on gelatin-methacryloyl and gellan gum for cartilage bioprinting. *Biofabrication*, 8, 3, 035003, 2016.
45. Lee, V., Singh, G., Trasatti, J.P., Bjornsson, C., Xu, X., Tran, T.N., Yoo, S.-S., Dai, G., Karande, P., Design and fabrication of human skin by three-dimensional bioprinting. *Tissue Eng. Part C Methods*, 20, 6, 473–484, 2014.
46. Kundu, J., Shim, J.H., Jang, J., Kim, S.W., Cho, D.W., An additive manufacturing-based PCL–alginate–chondrocyte bioprinted scaffold for cartilage tissue engineering. *J. Tissue Eng. Regen. Med.*, 9, 11, 1286–1297, 2015.
47. Apelgren, P., Amoroso, M., Lindahl, A., Brantsing, C., Rotter, N., Gatenholm, P., Kölby, L., Chondrocytes and stem cells in 3D-bioprinted structures create human cartilage *in vivo*. *PloS One*, 12, 12, e0189428, 2017.
48. Orloff, N.D., Truong, C., Cira, N., Koo, S., Hamilton, A., Choi, S., Wu, V., Riedel-Kruse, I.H., Integrated bioprinting and imaging for scalable, networkable desktop experimentation. *RSC Adv.*, 4, 65, 34721–34728, 2014.
49. Li, Z. and Kawashita, M., Current progress in inorganic artificial biomaterials. *J. Artif. Organs*, 14, 163–170, 2011.
50. Yuan, S. and Rigor, R., Structure and function of exchange microvessels, in: *Regulation of Endothelial Barrier Function*, D.N. Granger, and J.P. Granger, (Eds.), pp. 1–13, Morgan & Claypool Life Sciences, San Rafael, 2010.
51. Lawanyashri, M., Balusamy, B., Subha, S., Threshold-based workload control for an under-utilized virtual machine in cloud computing. *Int. J. Intell. Eng. Syst.*, 9, 4, 234–241, 2016.
52. Jia, W., Gungor-Ozkerim, P.S., Zhang, Y.S., Yue, K., Zhu, K., Liu, W., Pi, Q., Byambaa, B., Dokmeci, M.R., Shin, S.R., Direct 3D bioprinting of perfusable vascular constructs using a blend bioink. *Biomaterials*, 106, 58–68, 2016.

53. Kolesky, D.B., Homan, K.A., Skylar-Scott, M.A., Lewis, J.A., Three-dimensional bioprinting of thick vascularized tissues. *Proc. Natl. Acad. Sci.*, 113, 12, 3179–3184, 2016.
54. Lawanyashri, M., Balusamy, B., Subha, S., Energy-aware hybrid fruitfly optimization for load balancing in cloud environments for EHR applications. *Inform. Med. Unlocked*, 8, 42–50, 2017.
55. Pati, F., Jang, J., Ha, D.-H., Won Kim, S., Rhie, J.-W., Shim, J.-H., Kim, D.-H., Cho, D.-W., Printing three-dimensional tissue analogues with decellularized extracellular matrix bioink. *Nat. Commun.*, 5, 1, 3935, 2014.
56. Binder, K.W., Zhao, W., Aboushwareb, T., Dice, D., Atala, A., Yoo, J.J., *In situ* bioprinting of the skin for burns. *J. Am. Coll. Surg.*, 211, 3, S76, 2010.
57. Skardal, A., Mack, D., Kapetanovic, E., Atala, A., Jackson, J.D., Yoo, J., Soker, S., Bioprinted amniotic fluid-derived stem cells accelerate healing of large skin wounds. *Stem Cells Transl. Med.*, 1, 11, 792–802, 2012.
58. Lawanya Shri, M., Ganga Devi, E., Balusamy, B., Kadry, S., Misra, S., Odusami, M., A fuzzy based hybrid firefly optimization technique for load balancing in cloud datacenters, in: *Innovations in Bio-Inspired Computing and Applications: Proceedings of the 9th International Conference on Innovations in Bio-Inspired Computing and Applications (IBICA 2018)*, Kochi, India, December 17-19, 2018, vol. 9, pp. 463–473, Springer International Publishing, 2019.
59. Williams, D.F., On the mechanisms of biocompatibility. *Biomaterials*, 29, 20, 2941–2953, 2008.
60. West, J.L. and Hubbell, J.A., Polymeric biomaterials with degradation sites for proteases involved in cell migration. *Macromolecules*, 32, 1, 241–244, 1999.

# 3

# Design and Development of IoT Devices: Methods, Tools and Technologies

**Akash Kumar, Sachin Abhay Kumar, Richa Singh, Shivam Maloo, Dishant Rathi and K. Santhi***

*School of Information Technology and Engineering, Vellore Institute of Technology, Vellore, India*

***Abstract***

The design and development of IoT devices for 3D bioprinting is an emerging field that combines the latest advances in Internet of Things (IoT) technology with the cutting-edge field of 3D bioprinting. It provides an overview of the methodology, tools, and techniques used in the development of IoT devices for 3D bioprinting. The chapter begins with an introduction to the concepts of IoT and 3D bioprinting, and how they are integrated to create IoT devices for 3D bioprinting. The chapter then discusses the methodology used in the development of these devices, including the importance of understanding the specific requirements of the bioprinting process and the need for a multidisciplinary approach that combines hardware and software design, data processing, power optimization, quality assurance, and cybersecurity. Next, the chapter explores the tools that are commonly used in the development of IoT devices for 3D bioprinting, including sensors, microcontrollers, and communication protocols also examines the role of data analytics and machine learning in the development of these devices, and how they can be used to optimize the printing process and improve the accuracy and reliability of the printed tissues. Finally, the chapter discusses the various techniques used in the development of IoT devices for 3D bioprinting, including rapid prototyping, simulation, and testing. Also examines the importance of user interface design and ease of use in the development of these devices, particularly in medical applications where the devices may be used by non-technical personnel. Overall, this book chapter provides a comprehensive overview of the methodology, tools, and techniques used in the development of IoT devices for 3D bioprinting. It is a valuable resource for researchers, engineers, and professionals working in the fields of

**Corresponding author*: ksanthi@vit.ac.in

E. Gangadevi, M. Lawanya Shri, Rajesh Kumar Dhanaraj and Balamurugan Balusamy (eds.) Computational Intelligence in Bioprinting, (39–52) © 2024 Scrivener Publishing LLC

IoT and 3D bioprinting, and for anyone interested in the latest advances in this exciting and rapidly evolving field.

***Keywords:*** 3D bioprinting, sensors, IoT devices, cybersecurity, rapid prototyping

## 3.1 Introduction to IoT Devices and 3D Bioprinting

The internet of things (IoT) has revolutionized the way we interact with technology. It has enabled us to connect our devices to the internet and exchange data with other devices. This has opened up a world of possibilities, and one of the most promising is the use of IoT in 3D bioprinting.

3D bioprinting is a relatively new technology that has the potential to revolutionize medicine. It involves the use of 3D printers to create living tissues and organs that can be used for transplantation, drug testing, and other medical applications. By combining IoT with 3D bioprinting, we can create devices that are capable of monitoring and adjusting the environment in which the tissues are printed, leading to better quality and more consistent results.

The field of 3D bioprinting has shown great promise in the creation of complex, functional tissues and organs for transplantation and regenerative medicine. However, the printing process requires precise control over various parameters such as temperature, humidity, and oxygen levels, which can be difficult to achieve and maintain. This is where the Internet of Things (IoT) technology comes into play.

IoT devices can be used to monitor and control the printing environment in real-time, enabling precise control over the printing process. These devices can collect data from sensors and transmit it to other devices on the network, allowing for remote monitoring and control. The combination of IoT technology with 3D bioprinting has the potential to revolutionize the field of tissue engineering and regenerative medicine.

In this book chapter, we will explore the design and development of IoT devices for 3D bioprinting, including the methods, tools, and techniques used in this field. We will also discuss case studies of IoT devices for 3D bioprinting and the future directions in this field.

## 3.2 Methodology for Designing IoT Devices for 3D Bioprinting

The design and development of IoT devices for 3D bioprinting requires a structured approach that takes into account the unique challenges of this field. The following steps can be used to design an IoT device for 3D bioprinting:

1. Define the Problem: The first step in designing an IoT device for 3D bioprinting is to define the problem that the device is meant to solve. This could be anything from controlling the temperature and humidity of the printing environment to monitoring the quality of the printed tissue. It is important to have a clear understanding of the problem to be solved before proceeding with the design process.
2. Identify the Requirements: Once the problem has been defined, the next step is to identify the requirements for the IoT device. This includes the sensors that will be used, the data communication protocol, the power source, and the form factor. The requirements will vary depending on the problem to be solved, and it is important to carefully consider each requirement to ensure that the device will function as intended [2].
3. Develop a Prototype: With the requirements identified, the next step is to develop a prototype of the IoT device. This involves selecting the components, designing the circuit board, and creating a 3D model of the device. The prototype will allow designers to test the functionality of the device before moving on to the final design.
4. Test the Prototype: After the prototype has been developed, it is important to test it in a real-world environment. This involves testing the sensors, the communication protocol, and the power source to ensure that they are functioning correctly. Testing the prototype is a critical step in the design process, as it allows designers to identify any issues with the device and make adjustments before moving on to the final design.
5. Refine the Design: Based on the results of the prototype testing, the design can be refined to improve its performance. This may involve selecting different components or adjusting the placement of the sensors. Refining the design is an iterative process that may require multiple rounds of testing and adjustment.
6. Manufacture the Device: Once the design has been refined, the IoT device can be manufactured. This involves producing the circuit board, assembling the components, and testing the final product. Manufacturing the device requires careful attention to detail to ensure that the final product meets the design specifications.

7. Deploy and Monitor the Device: The final step in designing an IoT device for 3D bioprinting is to deploy it in the field and monitor its performance. This involves collecting data from the device and analyzing it to ensure that it is functioning correctly. Monitoring the device is an ongoing process that allows designers to identify any issues with the device and make adjustments as needed.

## 3.3 Additional Considerations in IoT Device Design for 3D Bioprinting

1. Safety: When designing an IoT device for 3D bioprinting, safety should be a primary consideration. The device should be designed to minimize the risk of contamination and ensure that the printing process is safe for both the operator and the patient.
2. Scalability: In some cases, the IoT device may need to be scaled up to accommodate larger printing environments or higher volumes of tissue production. Designers should consider scalability when developing the initial prototype to ensure that the device can be easily modified or replicated for larger applications [5].
3. Data Security: The IoT device will collect and transmit sensitive data related to the printing process. Designers should ensure that the device is designed with data security in mind to prevent unauthorized access to this information.
4. User Experience: The IoT device should be designed with the user in mind to ensure that it is easy to use and understand. This may involve incorporating user-friendly interfaces or simplifying the design to make it more intuitive [5].
5. Integration with Existing Systems: The IoT device may need to be integrated with existing 3D bioprinting systems or other IoT devices. Designers should consider how the device will be integrated into the existing infrastructure and ensure that it is compatible with other devices and software.

Overall, the methodology for designing IoT devices for 3D bioprinting involves a careful and iterative process that takes into account the specific requirements of the problem to be solved, as well as important considerations such as safety, scalability, data security, user experience, and integration with existing systems.

One key aspect of designing IoT devices for 3D bioprinting is selecting the appropriate sensors and communication protocol. For example, temperature and humidity sensors may be necessary to control the printing environment, while sensors for monitoring the quality of the printed tissue may also be needed. In addition, the communication protocol must be selected based on factors such as the range and reliability of the signal, as well as any potential interference from other devices in the environment.

Another important consideration is the power source for the IoT device. The device may need to be battery-powered or connected to a power source, depending on the specific application. Designers must carefully consider the power requirements of the device to ensure that it will function as intended.

The form factor of the IoT device is also an important consideration. The device must be designed to fit within the printing environment, while also being durable enough to withstand the stresses of the printing process. Designers may need to consider factors such as the material used for the device and the physical size and shape of the device.

The development of a prototype is a critical step in the design process, as it allows designers to test the functionality of the device in a real-world environment. This can involve testing the sensors, communication protocol, and power source to ensure that they are functioning correctly. Testing the prototype may require multiple rounds of adjustments and modifications to improve the performance of the device.

Once the design has been refined and the prototype has been tested, the device can be manufactured. This involves producing the circuit board, assembling the components, and testing the final product. Careful attention to detail is required during the manufacturing process to ensure that the final product meets the design specifications.

The final step in designing an IoT device for 3D bioprinting is to deploy it in the field and monitor its performance. This involves collecting data from the device and analyzing it to ensure that it is functioning correctly. Monitoring the device is an ongoing process that allows designers to identify any issues with the device and make adjustments as needed.

In conclusion, the methodology for designing IoT devices for 3D bioprinting involves a careful and iterative process that takes into account the specific requirements of the problem to be solved, as well as important considerations such as safety, scalability, data security, user experience, and integration with existing systems. By following this methodology, designers can create IoT devices that are effective, reliable, and safe for use in 3D bioprinting applications.

## 3.4 Tools for Developing IoT Devices for 3D Bioprinting

Developing IoT devices for 3D bioprinting requires the use of specialized tools and technologies. These tools range from hardware components and development boards to software and programming languages. In this section, we will explore some of the most commonly used tools for developing IoT devices for 3D bioprinting.

### 3.4.1 Microcontrollers and Development Boards

Microcontrollers are a fundamental component of IoT devices. They are responsible for processing data, controlling sensors and actuators, and communicating with other devices. The most popular microcontrollers for IoT devices are the Arduino and the Raspberry Pi. The Arduino is a simple microcontroller board that is easy to program, while the Raspberry Pi is a more powerful single-board computer that can run a full operating system.

### 3.4.2 Sensors and Actuators

Sensors and actuators are essential components of IoT devices. Sensors are used to detect changes in the environment, such as temperature, humidity, and pressure, while actuators are used to control devices, such as motors and valves. Common sensors used in 3D bioprinting applications include temperature sensors, humidity sensors, pressure sensors, and proximity sensors [3].

### 3.4.3 Communication Protocols

IoT devices need to communicate with each other and with other systems. There are several communication protocols that can be used for this purpose, including WiFi, Bluetooth, Zigbee, and LoRa. The choice of communication protocol will depend on the specific requirements of the application, such as range, bandwidth, and power consumption.

### 3.4.4 Software Development Kits

Software development kits (SDKs) provide developers with tools and resources for building software applications for IoT devices. These SDKs typically include programming languages, libraries, and development tools. Popular IoT SDKs include the Arduino IDE, the Raspberry Pi OS, and the Amazon Web Services IoT SDK [3].

### 3.4.5 Cloud Platforms

Cloud platforms provide a way to store and process data from IoT devices in the cloud [8]. This allows developers to analyze data from multiple devices [9] and create more advanced applications. Popular cloud platforms for IoT devices [10] include Amazon Web Services, Microsoft Azure, and Google Cloud [7].

### 3.4.6 3D Printing Software

3D bioprinting requires specialized software for designing and printing tissues. Popular 3D printing software for bioprinting includes Autodesk's Netfabb and Ultimaker's Cura.

### 3.4.7 CAD Software

CAD software is used to design 3D models for printing. Popular CAD software for 3D bioprinting includes Autodesk's Fusion 360 and SolidWorks.

### 3.4.8 Simulation Software

Simulation software can be used to test the performance of IoT devices before they are deployed in the field. This software can simulate various scenarios and environments to ensure that the device will function correctly. Popular simulation software for IoT devices includes ANSYS, MATLAB, and Simulink.

### 3.4.9 Data Analytics Tools

Data analytics tools can be used to analyze the data collected from IoT devices. These tools can identify patterns and trends in the data, which can be used to optimize the performance of the device. Popular data analytics tools for IoT devices include Apache Spark and Hadoop.

### 3.4.10 Cybersecurity Tools

Cybersecurity is a critical consideration for IoT devices, as they are often connected to the internet and can be vulnerable to cyberattacks. There are several cybersecurity tools available for IoT devices, such as firewalls, intrusion detection systems, and encryption software.

In conclusion, developing IoT devices for 3D bioprinting requires the use of specialized tools and technologies. These tools range from microcontrollers and development boards to software and programming languages. By using these tools, developers can create effective and reliable IoT devices for 3D bioprinting applications.

## 3.5 Techniques for Developing IoT Devices for 3D Bioprinting

Developing IoT devices for 3D bioprinting requires the use of various techniques to ensure that the devices are reliable, efficient, and effective. These techniques range from hardware design to software development and testing. In this section, we will explore some of the most commonly used techniques for developing IoT devices for 3D bioprinting [6].

### 3.5.1 Agile Development

Agile development is a methodology for software development that emphasizes flexibility and collaboration. This methodology is particularly useful for IoT devices because it allows developers to quickly adapt to changing requirements and to test their devices in real-world scenarios. Agile development involves continuous iteration, feedback, and improvement [4].

### 3.5.2 Rapid Prototyping

Rapid prototyping is a technique for quickly developing and testing prototypes of IoT devices. This technique involves creating a basic prototype of the device and then testing it to identify any flaws or issues. Once the issues are identified, the prototype can be refined and improved. Rapid prototyping is particularly useful for IoT devices because it allows developers to quickly iterate and improve their designs.

### 3.5.3 Test-Driven Development

Test-driven development (TDD) is a software development technique that involves writing tests before writing the code. This technique ensures that the code is reliable, efficient, and effective. TDD is particularly useful for IoT devices because it allows developers to quickly identify and fix issues with the code.

### 3.5.4 Continuous Integration

Continuous integration (CI) is a technique for automatically building, testing, and deploying code changes. This technique ensures that the code is reliable and that any issues are identified and fixed quickly. CI is particularly useful for IoT devices because it allows developers to quickly iterate and improve their devices.

### 3.5.5 Modular Design

Modular design is a technique for designing IoT devices that consists of multiple independent modules. Each module is responsible for a specific function, such as data processing, communication, or sensing. Modular design is particularly useful for IoT devices because it allows developers to quickly replace or upgrade individual modules without affecting the rest of the device.

### 3.5.6 Power Optimization

Power optimization is a technique for optimizing the power consumption of IoT devices [8, 9]. This technique involves reducing the power consumption of individual components by optimizing their usage or by using low-power components. Power optimization is particularly important for IoT devices because they are often battery-powered and need to operate for long periods of time.

### 3.5.7 Data Processing Techniques

Data processing techniques are used to process and analyze the data collected from IoT devices. These techniques include machine learning, data mining, and statistical analysis. Machine learning, for example, can be used to identify patterns and trends in the data, while statistical analysis can be used to identify anomalies [10].

### 3.5.8 Quality Assurance

Quality assurance is a technique for ensuring that IoT devices are reliable, efficient, and effective. This technique involves testing the devices in a variety of scenarios and environments to ensure that they perform correctly. Quality assurance is particularly important for IoT devices because they are often used in critical applications, such as medical devices.

### 3.5.9 Cybersecurity Techniques

Cybersecurity techniques are used to ensure that IoT devices are secure and protected from cyberattacks. These techniques include encryption, firewalls, intrusion detection systems, and access controls. Cybersecurity is particularly important for IoT devices because they are often connected to the internet and can be vulnerable to cyberattacks.

### 3.5.10 Standardization

Standardization is a technique for ensuring that IoT devices are compatible with other devices and systems. Standardization involves defining common protocols and interfaces that can be used by different devices. Standardization is particularly important for IoT devices because they often need to communicate with other devices and systems.

In conclusion, developing IoT devices for 3D bioprinting requires the use of various techniques to ensure that the devices are reliable, efficient, and effective. These techniques range from rapid prototyping and modular design to power optimization and cybersecurity. By using these techniques, developers can create IoT devices that meet the specific requirements of 3D bioprinting and provide accurate and reliable data for the printing process.

One important aspect of developing IoT devices for 3D bioprinting is ensuring that the devices are compatible with the specific bioprinting process being used. For example, if the bioprinting process requires precise temperature control, the IoT device must be able to accurately measure and control temperature. Similarly, if the bioprinting process requires monitoring of specific parameters such as pH or oxygen levels, the IoT device must be able to accurately measure and report those parameters.

Another important aspect of developing IoT devices for 3D bioprinting is ensuring that the devices are user-friendly and easy to operate. This is particularly important in medical applications where the devices may be used by non-technical personnel. To achieve this, developers must consider the user interface design and the ease of use of the device.

Overall, the development of IoT devices for 3D bioprinting requires a multidisciplinary approach that combines hardware and software design, data processing, power optimization, quality assurance, and cybersecurity. By using these techniques, developers can create IoT devices that are reliable, efficient, and effective, and that provide accurate and reliable data for the bioprinting process. With the continued growth and development of the IoT industry, it is likely that we will see many more advances in the design and development of IoT devices for 3D bioprinting in the years to come.

## 3.6 Case Studies of IoT Devices for 3D Bioprinting

There are a growing number of IoT devices that are being developed for 3D bioprinting. In this chapter, we will explore some case studies of IoT devices for 3D bioprinting, including:

The BioScaffolder: The BioScaffolder is an IoT device that is designed to control the printing environment during 3D bioprinting. It includes sensors for monitoring temperature, humidity, and pH, as well as actuators for controlling these parameters.

The Biobot: The Biobot is an IoT device that is designed to monitor and control the printing process in real-time. It includes sensors for monitoring temperature, pH, and oxygen levels, as well as a communication module for transmitting data to other devices on the network.

The 3D-Bioplotter: The 3D-Bioplotter is a 3D printer that is designed specifically for printing living tissues and organs. It includes an IoT device that is used to control the printing environment and ensure that the tissues are printed under optimal conditions.

## 3.7 Future Directions in IoT Devices for 3D Bioprinting

The field of IoT devices for 3D bioprinting is rapidly evolving, and there are many exciting opportunities for future research and development. In this chapter, we will explore some of the future directions in this field, including:

Integration with AI and machine learning: IoT devices for 3D bioprinting can be integrated with AI and machine learning algorithms to enable real-time monitoring and control of the printing process [1].

Nanotechnology and bioinks: Advances in nanotechnology and bioinks are opening up new possibilities for 3D bioprinting, and IoT devices can be used to monitor and control the printing process at the nanoscale.

Personalized medicine: IoT devices for 3D bioprinting can be used to create personalized tissues and organs that are tailored to the specific needs of individual patients.

Integration with robotics: Robotics technology can be used to automate the 3D bioprinting process, and IoT devices can be used to monitor and control the printing process in real-time.

## 3.8 Conclusion

IoT devices are transforming the field of 3D bioprinting, enabling real-time monitoring and control of the printing environment. The combination of IoT technology with 3D bioprinting has the potential to revolutionize the field of tissue engineering and regenerative medicine by enabling the creation of complex, functional tissues and organs.

In this book chapter, we have explored the design and development of IoT devices for 3D bioprinting, including the methods, tools, and techniques used in this field. We have also discussed some case studies of IoT devices for 3D bioprinting and the future directions in this field.

As the field of IoT devices for 3D bioprinting continues to evolve, it is clear that there are many exciting opportunities for future research and development. By combining IoT technology with 3D bioprinting, we can create new possibilities for tissue engineering and regenerative medicine, enabling the creation of complex, functional tissues and organs that can improve the quality of life for patients around the world.

## References

1. https://www.nano-di.com/resources/blog/2019-how-to-design-an-iot-device-for-3d-printing
2. https://www.startertutorials.com/blog/iot-design-methodology.html
3. https://towardsdatascience.com/a-comprehensive-guide-to-start-building-an-iot-product-ba32dfb91c7a
4. https://www.scnsoft.com/blog/software-development-models
5. Fatima, S., Haleem, A., Bahl, S., Javaid, M., Mahla, S.K., Singh, S., Exploring the significant applications of internet of things (IoT) with 3D printing using advanced materials in medical field. *Mater. Today Proc.*, 45, 4844–4851, 2021.

6. Shahrubudin, N., Lee, T.C., Ramlan, R., An overview on 3D printing technology: Technological, materials, and applications. *Proc. Manuf.*, 35, 1286–1296, 2019.
7. Murthy, C.V.B., Shri, M.L., Kadry, S., Lim, S., Blockchain based cloud computing: Architecture and research challenges. *IEEE Access*, 8, 205190–205205, 2020.
8. Lawanya Shri, M., Ganga Devi, E., Balusamy, B., Kadry, S., Misra, S., Odusami, M., A fuzzy based hybrid firefly optimization technique for load balancing in cloud datacenters, in: *Innovations in Bio-Inspired Computing and Applications: Proceedings of the 9th International Conference on Innovations in Bio-Inspired Computing and Applications (IBICA 2018)*, Kochi, India, December 17-19, 2018, vol. 9, Springer International Publishing, pp. 463–473, 2019.
9. Lawanyashri, M., Balusamy, B., Subha, S., Energy-aware hybrid fruitfly optimization for load balancing in cloud environments for EHR applications. *Inform. Med. Unlocked*, 8, 42–50, 2017.
10. Lawanyashri, M., Balusamy, B., Subha, S., Threshold-based workload control for an under-utilized virtual machine in cloud computing. *Int. J. Intell. Eng. Syst.*, 9, 4, 234–241, 2016.

4

# AI-Based AR/VR Models in Biomedical Sustainable Industry 4.0

**Kuldeep Singh Kaswan[1]*, Jagjit Singh Dhatterwal[2], Rudra Pratap Ojha[3], Balamurugan Balusamy[4] and Gangadevi E.[5]**

*[1]School of Computing Science and Engineering, Galgotias University, Greater Noida, India*
*[2]Department of Artificial Intelligence & Data Science, Koneru Lakshmaiah Education Foundation, Vaddeswaram, Andhra Pradesh, India*
*[3]Department of Computer Science & Engineering, GL Bajaj Institute of Technology and Management, Greater Noida, India*
*[4]Computer Science and Engineering, Shiv Nadar University, Delhi, India*
*[5]Department of Computer Science, Loyola College, Chennai, India*

## *Abstract*
The biomedical field has witnessed significant advancements in recent years, with the emergence of technologies such as Augmented Reality (AR), Virtual Reality (VR), and Artificial Intelligence (AI). This abstract explores the integration of these technologies in the sustainable Industry 4.0 framework, specifically focusing on their applications in bioprinting services. Bioprinting, an innovative technique that enables the fabrication of three-dimensional biological structures, has immense potential in various biomedical applications, including tissue engineering, regenerative medicine, and drug development. However, the complexity of bioprinting processes, coupled with the need for precision and efficiency, demands advanced tools and methodologies. The integration of AI-based algorithms in AR/VR models holds tremendous promise for enhancing bioprinting services in terms of accuracy, productivity, and sustainability. By utilizing AI, these models can analyze vast amounts of data, including patient-specific medical information, to generate optimized bioprinting patterns and strategies. AI algorithms can also facilitate real-time monitoring of the bioprinting process, enabling adaptive control and reducing errors. Furthermore, the integration of AR/VR technologies in the bioprinting workflow allows for immersive visualization and interaction

**Corresponding author*: kaswankuldeep@gmail.com

E. Gangadevi, M. Lawanya Shri, Rajesh Kumar Dhanaraj and Balamurugan Balusamy (eds.) Computational Intelligence in Bioprinting, (53–78) © 2024 Scrivener Publishing LLC

with the bioprinted structures. This enhances the design and validation processes, enabling researchers and clinicians to evaluate the functionality and compatibility of the printed tissues or organs before implementation. In the context of Industry 4.0, the utilization of AI-based AR/VR models in bioprinting services can lead to sustainable advancements. By optimizing the printing process, reducing material waste, and minimizing errors, these technologies contribute to the overall sustainability of bioprinting services. Moreover, the integration of AI algorithms allows for continuous learning and improvement, enhancing the efficiency and quality of the bioprinting workflow. This abstract highlights the potential of AI-based AR/VR models in the biomedical field, specifically in the context of sustainable Industry 4.0 for bioprinting services. The fusion of these technologies can revolutionize the field of bioprinting, leading to the development of personalized and functional tissues and organs, and ultimately transforming the landscape of regenerative medicine and health care as a whole.

***Keywords:*** Augmented reality, VR, bioprinting

## 4.1 Introduction

The integration of artificial intelligence (AI) with augmented reality (AR) and virtual reality (VR) technologies has the potential to revolutionize the biomedical field, particularly in the context of sustainable Industry 4.0 for bioprinting services. Bioprinting, which involves the fabrication of three-dimensional (3D) tissues and organs using bioinks and additive manufacturing techniques, holds great promise for advancements in regenerative medicine, drug discovery, and personalized health care. By incorporating AI into AR/VR models, the capabilities and sustainability of bioprinting services can be significantly enhanced.

AR/VR technologies provide immersive and interactive experiences that allow users to visualize, manipulate, and interact with virtual objects in real-time. When combined with AI algorithms, these technologies can optimize the bioprinting process and improve overall sustainability [1].

Researchers in recent years have been observing both augmented reality and Industrial 4.0. It shows that few research efforts have focused on the potential of augmented reality (AR) in Industry 4.0 contexts.

Optimized Bioprinting Process: AI algorithms can analyze large datasets and experimental data to identify patterns and optimize the bioprinting process. By considering factors such as bioink composition, cell viability, printing parameters, and structural integrity, AI can assist in real-time decision-making to achieve precise and efficient bioprinting outcomes. AR/VR models can then visualize and simulate the optimized printing process, enabling users to understand and validate the proposed changes [2].

Predictive Modeling and Simulation: AI-driven AR/VR models can generate predictive models and simulations to anticipate the behavior and functionality of bioprinted tissues and organs. By training on existing datasets, AI algorithms can generate virtual representations of complex biological structures, predict their mechanical properties, and simulate the effects of different printing strategies. This allows for the assessment of printability, structural integrity, and functional performance before the physical bioprinting process, saving time, resources, and reducing waste.

AR/VR platforms powered by AI can facilitate remote collaboration and training among multidisciplinary teams involved in bioprinting services. Researchers, clinicians, and engineers can virtually collaborate and share expertise, regardless of their geographical locations. AI algorithms can support real-time data analysis, provide feedback, and assist in decision-making during collaborative sessions. Additionally, AR/VR models can serve as training tools, enabling novices to learn and practice bioprinting techniques in a virtual environment before performing them in the laboratory. These statistics suggest that the majority of the literature on augmented reality is devoted to either theoretical comparison between augmented reality and virtual reality or mixed reality, or to more broad studies that point to augmented reality's potential usefulness in fields like design, education, engineering, and visualization [3].

This chapter aims to inform readers on how augmented reality (AR) may be used in a manufacturing environment, including its potential benefits and downsides.

Authors draw on their own experiments and existing material to make observations regarding augmented reality (AR) and its use within an Industry 4.0 setting. Scopus has been designated as the primary database from which to pull research articles for this investigation. Since there are so many papers written on AR and Industry 4.0, the major criteria for revealing the finest AR industrial applications has been narrowed down to keyword combinations using both "Augmented Reality" and another word. For this purpose, we have utilized search terms like "maintenance," "training," "manufacturing," and "medical" to narrow down the available material [4].

The paper's rational progression is shown in Figure 4.1. In this article, we conduct the first comprehensive examination of the current level of augmented reality technology, revealing the top-tier gear and software that can be used in every conceivable AR setting. Following such investigation, industrial applications are analyzed while considering not just hardware/software constraints but also those imposed by real-world working situations. The topic of whether or not AR is an established technology for

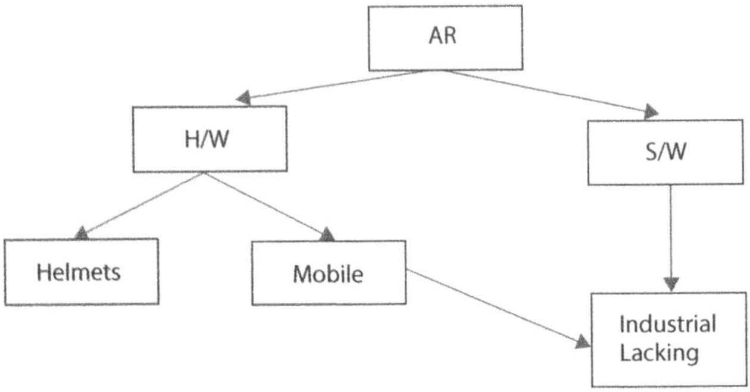


**Figure 4.1** Structure of industrial application.

commercial use is resolved in the last part, which also includes recommendations for adapting AR to AR4.0 and beyond.

With this context, the paper is structured as follows: The rest of the paper is organized as follows after the introduction: we will go through the specifics of augmented reality. The third section provides an overview of existing augmented reality applications, the fourth describes how the authors think the technology may be enhanced to achieve Technological Readiness Level 9, and the fifth provides a conclusion and summary.

## 4.2 Mixed Augmented Reality

There are a number of packages and tools out there now that make creating AR and VR applications more simpler. Mixed realities, often known as hybrid realities, include Augmented Reality (AR), Augmented Virtuality (AV), and Virtual Reality (VR) [4, 5]. The first is the use of transparent head-mounted displays to incorporate virtual items into the physical world. This technology facilitates the merging of the virtual and the physical, enhancing our ability to see the world as it really [5]. The second concept is incorporating physical items into a digital setting. An in-depth explanation of AV and its maintenance applications may be found [6]. Virtual reality (VR) is an artificial environment where the user experiences events and objects from the first person perspective in a purely digital setting. Virtual reality headsets (HMDs) and other immersive equipment (Oculus Rift, PlayStation VR) are required for usage. The notion of "Mixed Reality" within this context centers on the interaction between the physical and digital worlds. Blending virtual elements into the physical world and vice versa, Mixed Reality technology fill the threshold between the two realms.

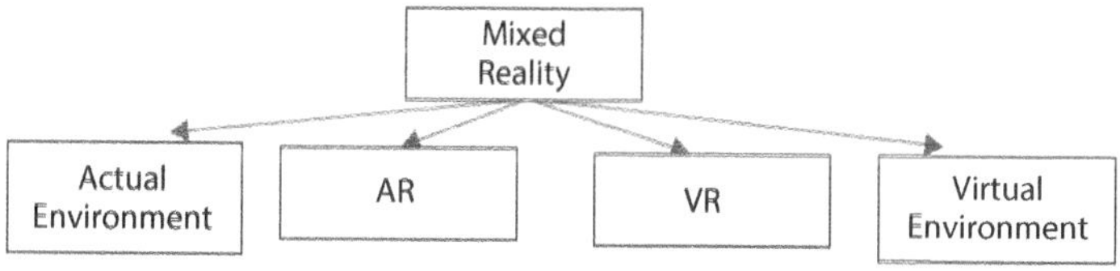


**Figure 4.2** Structure of mixed reality.

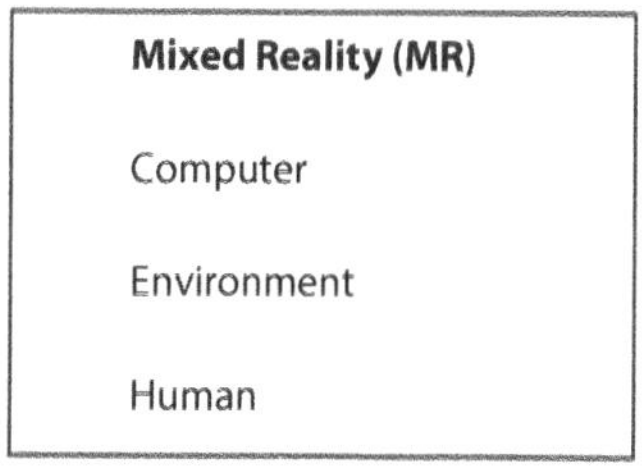


**Figure 4.3** Human–computer–interaction.

The "Reality-Virtuality Continuum" described in the original work [7] is a significant addition to this field.

Both the "Real World" and "Virtual Reality" are included in this category [8]. It's possible to see the concepts of "Augmented Reality" and "Augmented Virtuality" as "transitional phases" between the (see Figure 4.2). By narrowing this chasm, we can provide the user a more lifelike picture of the virtual environment and better overall experiences (see Figure 4.3).

The software and hardware tools and devices created to enable AR are described in further depth in the next two sections.

### 4.2.1 SDK in Augmented Reality

There are a plethora of SDKs available, and an evaluation of these various virtual reality/augmented reality libraries is accessible [9]. In order to properly refer to the various reference systems (camera, markers, object) needed to implement immersive virtual reality, white and black markers are used by market-ready marker-based technologies like ARToolkit. More sophisticated technology, including a marker-less augmented reality method, may be used with other software available on the market, such as Vuforia [10]. In the second scenario, the external scene is used to determine the camera's direction and location, eliminating the need for a predetermined marker. Using pattern recognition [12], this library can convert a real-world 2D picture or 3D item into a marker, giving the end user more options. Figure 4.4 shows how Vuforia may be used to create an augmented

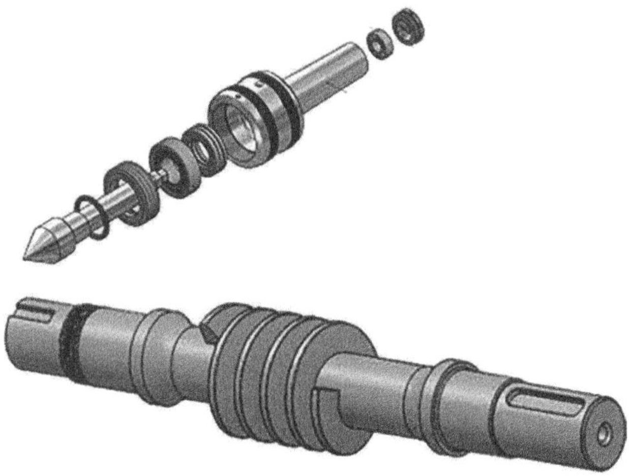

**Figure 4.4** SDK in augmented reality.

reality software, with a virtual picture referencing a physical controller device using marker-less technology.

To move augmented reality (AR) beyond the realm of academic curiosity and into practical use in industry, this approach is essential [13]. In execution of the process, for example, the item itself may serve as a marker, doing away with the need for separate markings or other cumbersome surveillance equipment. Both Google and Apple have released major new tools in the last several months, with the former promoting theirs as ARCore [14] and ARkit [15]. Both devices use the camera to keep an eye on their immediate environment. In the case of the former, Google Pixel phones are supported, along with an expanding number of other high-end gamma smartphones. It evolved from the experimental Project Tango. In contrast to ARCore, which works on all platforms, the ARkit is Apple-only. Shadows of 3D objects may be cast by these programs, and holograms may be fixed to existing maps of the horizontal environment (updates are now exploring the potential of also working with vertical planes). Image processing libraries like OpenCV are crucial for image tracking because they can read camera data and apply filters and mathematical algorithms to identify patterns. Software like Unity and Unreal Engine makes it simple to organize and use several libraries for image processing. Unreal Engine is still a newbie to the augmented reality/virtual reality business, but Unity developers have more experience designing efficient and effective integration solutions. Unity may be thought of as a container that facilitates the simultaneous execution of several scripts and the creation of the final environment in which an augmented reality app will operate [11]. This utility allows for the compilation of applications for several platforms, including Windows, Mac OS, Linux, iOS, and Android. The main benefit and drawback of the program is that it is well-suited for general-purpose computer graphic applications, such as a real-time render engine that allows the user to interact with the AR or VR in the most natural way possible, but it is not suitable for mechanical and industrial applications. Unity is not a CAD (computer-aided design) program. Nevertheless, you are unable to modify complicated models or deal with parametric items (as even the most basic 3D CAD can today). Even if the software isn't quite there yet, it's great for gathering models, creating scenarios that include interactions among objects, and putting together a final application. This assertion is supported by the authors' experimental findings. The ARdroid app, an augmented reality software for smartphones based on the Unity platform, was created in order to carry out an efficient evaluation. As long as Unity's Astronaut marker is present, OBJ-formatted CAD models may be overlaid on top of a smartphone's camera feed. Everything functions normally when a small OBJ model is loaded (1 MB), but as can be seen in Figure 4.5, problems with visualization arise when larger models are loaded (e.g., >3 MB). The top picture compares the high-quality OBJ version of the identical CAD model to the low-quality STL version. The low-resolution model is just 142 kb in size, which is much less than the high-resolution version. The results are self-explanatory, despite the fact that the visualization has flaws (the coding and other surface errors conceal a piece of the heavier model).

**Figure 4.5** CAD structure.

These issues are linked to Unity's polygon processing capacity, which has a hard limit.

### 4.2.2 Application Scope of Augmented Reality

Which hardware to utilize for augmented reality depends mostly on the scope of the final application. Head Mounted Displays (HMDs) are crucial when thinking about the hardware developed to satisfy augmented reality's needs. Video see-through (VST) and optical see-through (OST) HMDs are the two most common types (OST). A camera that records the environment for further analysis is the basis of VST technology. In this case, digital icons or models are placed on top of the live-streaming footage. In contrast, OST technology accomplishes its goals by projecting synthetic virtual models onto semi-transparent lenses; in this setup, the user sees both the real world and the virtual model coexisting in a single field of view [18]. Contrarily, VST hardware produces an accurate representation of the physical environment on the screen [16]. Smartphone applications are excellent examples of this kind of technology. Separating immersive virtual reality hardware based on its processing unit yields two subsets: dedicated hardware machines, which have their own system is incorporated and processors to frame the real world, superimpose models, and visualize the final live broadcasts, and devices that need additional hardware, that must be attached via cables, Bluetooth, or Wi-Fi to work.

#### *4.2.2.1 Video Capabilities*

In the context of AI-based AR/VR models in the biomedical field and sustainable Industry 4.0 for bioprinting services, video capabilities through devices play a crucial role in enhancing the overall experience and functionality. Here's how video integration can be beneficial in this context:

Remote Collaboration and Consultation: Video capabilities enable remote collaboration and consultation among experts, researchers, and health care professionals involved in bioprinting services. Through AR/VR interfaces, individuals can share live video feeds of their workstations, bioprinting setups, or laboratory environments, allowing real-time discussions, feedback, and guidance regardless of geographical distances. This fosters collaboration, knowledge exchange, and expert support, leading to improved outcomes and accelerated progress in bioprinting research and services.

Real-time Monitoring and Feedback: Video feeds from devices can be integrated into AI-based AR/VR models to provide real-time monitoring and feedback during the bioprinting process. By streaming live video from cameras positioned within bioprinters or incubators, operators can visualize the printing process, monitor cell behavior, assess environmental conditions, and identify any deviations or abnormalities. AI algorithms can analyze these video feeds, provide automated feedback, and trigger alerts in case of potential issues, enabling prompt interventions and ensuring quality control.

Training and Education: Video integration in AI-based AR/VR models enhances training and education in bioprinting services. By capturing video recordings of expert bioprinting procedures, these recordings can be used as educational resources for trainees and novice users. Through AR/VR interfaces, trainees can view the recorded videos and observe the step-by-step process, techniques, and best practices. This visual guidance, combined with interactive virtual elements, facilitates skill development, knowledge transfer, and ensures consistency in bioprinting techniques across different operators.

#### *4.2.2.2 AR Toolkit Technology*

OST may take use of a toolkit of technologies made for assisting with AR applications. The projection methods, which are likely the beating core of OST HMDs, have received the most attention from academics and businesses [17]. The half-mirror technique is the simplest method for projecting pictures in HMD, although it has a small FOV(FoV). Convex or free-form shaped mirrors add complexity, but they help expand vision range. The Meta2 device, for instance, uses such a mirror to increase the viewing area. (FoV) and high display resolution. The waveguide grating is the cutting edge of optical technology, combining holographic and diffractive optics. Like a lens, the diffraction grating converges light onto a flat, transparent surface. The channel guides the reflected light upward, where it may be seen by the user. This method is implemented in Microsoft's HoloLens product line. A growing number of companies now provide their own augmented reality hardware. Although Google Glass was the pioneering product of its sort, Wuzix provided a selection of wearable devices, some of which could function without a host device. Meta2 and HoloLens are two instances of Microsoft's proprietary software. The Meta2 does not have any form of integrated computing capacity, unlike the Microsoft HoloLens and several varieties of the family Vuzix Glasses. To improve the user experience, Microsoft HoloLens makes advantage of the computing power supplied by an Intel Atom CPU, much as the Vuzix M300 Smart Glasses. In contrast to the Microsoft HoloLens, which can run any software independently, Meta2 headsets need an external computer in order to work. HoloLens is a wearable computer with a holographic display. Vuzix M300 Smart Glasses are Android-powered, so you may use them with any number of other mobile and eyewear applications. Microsoft's HoloLens employs waveguide-based holographic lenses, whereas devices like the Meta2 use convex mirrors to project images in front of the user's line of sight. The Meta2 method seems to be capable of guaranteeing a larger field of view than HoloLens, which might be useful in creating a more interactive environment. It's true that the experimenter will have a stereographic 3D vision experience; but, because to the device's poor FoV, the experience will be limited to what's in front of their face. The larger field of view of the Meta2 is useful in the field, but there is a risk that the user's eyesight may be obscured by holograms. The Vuzix M300 Smart Glasses, the company's crowning achievement to date, include a monocular design and a small screen that does not encompass the user's whole field of view.

One of the most noticeable differences between Meta2 and HoloLens is how it handles gestures. Meta2 outperforms HoloLens because the gestures it employs are more intuitive to utilize. After putting both gadgets through their paces, the authors concluded that the Meta2's method—which entails using the complete hand to control holograms and the fingertips to pick and move icons on the screen—was more effective. The motion necessary to utilize HoloLens is described in the literature as "blooming," after the unfolding of a flower's petals. For most individuals, this is an uncomfortable and unnatural action. The alternative technique of selecting in HoloLens, which uses a stylized "air tapping" finger motion, has also been criticized. One major problem is that good catching requires the use of both the index finger and the thumb. Most people's fingers are able to rotate, making it difficult to use the Hololens, which needs the index to be held straight out in front of the sensors. Last but not least, "air tapping" might be unpleasant in certain workplaces due to the user having to hold his arm in an unnatural position, front of his head. The Vuzix M300 Smart Glasses come with software for use with tablets and smartphones as controllers; this is useful since many mobile applications need a human-machine interface, such as a touchpad or a keyboard. The application works with iOS and Android smartphones.

#### *4.2.2.3 Quality of Tracking System*

The use of a tracking system in this analysis helps to highlight the distinctive qualities of the evaluated devices. The Oculus Rift has the same number of sensors (IR Led) that can follow the user across 360 degrees, but in a smaller area specified by 2.5 square meters, whereas the HTC Vive's infrared (IR) sensor can perform 360-degree tracking and cover an area of 4.5 x 4.5 m. The HoloLens, on the other hand, has no such restrictions since its sensor is continually mapping the playing environment. It's a significant plus that HoloLens doesn't need any special setup to work. Two "environment understanding cameras" on each side form the basis of its tracking system. With the help of the depth sensor and "environment understanding cameras," we can do both head tracking and environment identification. The latter feature lets the user specify where an object should be placed in three-dimensional space. Like HoloLens, Meta2's tracking abilities stem from a combination of cameras, depth sensors, and an inertial measurement unit (IMU). The data fusion is obtained by a method that is similar to SLAM (Simultaneous Localization and Mapping). Separate sensors, such as the Kinect Sensor, Leap Motion, and Occipital Structures Sensor 3D Scanner, are also obtainable in addition to these built-in devices. These devices are useless until linked to a computer, but once functional, they provide promising results at a low cost. Keep in mind that these tools enabled for early software development, even before integrated helmets were commonplace. Microsoft Kinect connects the two items despite their shared technology (webcam and infrared sensor). The gesture algorithms also make advantage of the information obtained by Leap Motion, a device devoted to hand recognition for object interaction. This device has the potential to recognize each of a user's fingers, allowing for the assignment of unique actions to each. Two cameras and three infrared (IR) sensors work together to make this possible. Leap Motion's precision up to 0.40 m from the sensor makes it a popular addition to existing helmets for enhancing user engagement with the augmented reality. Like the Structure Sensor, which can be joined to any number of devices through USB, the Structure Sensor is the first 3D sensor created with mobile devices, such as tablets and smartphones in mind. High-resolution, full-color 3D scans of objects or people may be taken in real time by linking the sensor to an appropriate mobile device. The Unity 3D software development kit (SDK) allows for the creation of augmented reality (AR) and virtual reality (VR) experiences. Unlike the HTC Vive, which requires significant computing resources and the installation and calibration of fixed lighthouses, the Structure Sensor device can provide positional monitoring with little effort.

If present trends in AR equipment development persist, the divide between wearable technology and the actual world will narrow. The concept of a digital twin will get even more attention when future technologies make it possible to switch reliably between the real item and its digital copy in real time. This is done to facilitate real-time alert management systems, enhance the security of human-robot interactions, and better inform the decisions of engineers working with real-world data [19].

## 4.3 AR Technology

This section discusses the advantages and disadvantages of using AR in the business world. The fundamental issue is that although existing AR technology offers high visualization performances, it lacks interactive capabilities, as will be more fully detailed below. These parts include assertions from the writers' point of view about AR, supported by examples and case studies the authors themselves have prepared [20].

### 4.3.1 High Level Augmented Reality

This section gives a brief summary of some of the applications of augmented reality that have been discussed in the academic literature.

The use of augmented reality (AR) to help with industrial maintenance is described in detail in a number of papers [21]. While the addition of text and animations depicting elected parliament methodologies can aid in the resolution of complex engineering challenges by inexperienced operators, the pictorial representation of virtual models [22] can be useful for identifying the right combination to operate on when dealing with complex products. Moreover, AR has the potential to facilitate the simplification of user/maintenance guides. Instead of relying on detailed drawings that might be difficult to interpret or miss altogether, a hands-on approach grounded in the real world may be more efficient and save time in this situation [23]. The research suggests that augmented reality's potential benefits extend to maintenance education as well. This is because virtual models may be used in place of the actual thing when training operators, removing the requirement for the trainer to be in the same room as the equipment and reducing the risk of injury. Remote collaborative possibilities are said to exist in the reference [24], where a remote operation center may aid an operator in complex operations by providing the appropriate animations to assist him or her in the field. Pick by vision is another intriguing notion that has been investigated in the literature to help operators decide which parts to pick up during assembly or logistics activities. The use of an AR "virtual eye bull" or "tunnel" to guide the user to the required part has also been proposed. The manufacturing environment is perfect for augmented reality applications due to the abundance of readily recognizable symbols. Outdoor apps, on the other hand, need the use of Global Positioning

System (GPS) and Inertial Measurement Units (IMU) to fix the position and point of view of the operator; exploring the unexplored real world may necessitate a heavy backpack with a compass, IMU, and GPS [25].

Medical situations are also described in considerable depth in the literature for the sake of testing and employing AR. Some of the many ways in which augmented reality (AR) might aid in surgical training include facilitating the precise positioning of needles and medical tools, improving visibility, and facilitating the alignment of CT images with the actual patient. To overcome these obstacles, researchers are developing augmented reality (AR) systems that can continue tracking even if a medical operator covers the field of view of cameras or markers, and AR glasses that can be worn by patients. The need for high visual contrast might be problematic in a real operating room, where the lights are on a dimmer switch at the doctors' discretion. Recent efforts in the field of augmented reality in medicine have focused on developing applications for OST HMDs to help surgeons [26].

The electronic industry can count on a substantial influx of cash from the entertainment sector. Mobile and home Nintendo entertainment system have long supported first-person and scene-augmentation games. Augmented reality (AR) may be viewed as a mature, if expensive, technology ready to be employed in games [27], since exact placement and excellent performance aren't needed here.

Many smartphone-based outdoor control systems exist to aid tourists by giving them a comprehensive history of the locations they visit. Well-known imagery, such a mountain silhouette or a building, may be followed in such a situation. Nevertheless, the accuracy requirements of such applications are far lower than those of the medical industry. Although high-performance airplanes often prioritize costly technology like heads-up displays (HUDs) with information for pilots, the car sector is considering augmented reality (AR) technologies like heads-up displays to assist the driver and offer guidance information on the windshield. To monitor your speed, RPMs, fuel level, settings, or knob position, you won't have to take your sights from the road [28].

Toy manufacturers often use this function to promote the contents of their package by overlaying pictures and text onto the actual environment, or to gain Augmented Business Cards that provide data in three dimensions. When a consumer is near a retail sign, certain augmented reality applications may deliver more details to their phone. As symbols and animations only need to be created once and then reused across many jobs, the ability to run augmented reality applications on smartphones or cheap devices is more important than having high refresh rates or excellent precision in location in these situations.

### 4.3.2 Limitations of Enhanced Image

Each of the several commonly produced projection technologies has its own set of advantages and disadvantages. Lightweight and small, the device's shape limits the field of vision (FoV) given by the half-mirror technique. A convex mirror-based technique is used to increase the field of view (e.g., 90 degrees in the Meta2 HMD). The combiner (the device upon which the images are projected) is made of plastic, therefore the resolution may change depending on how well it holds up over time. In addition, the FoV afforded by the waveguide is less than that afforded by the conventional mirror configuration. The dimness of the OST HMD is a major drawback. Because of this, the melding of the virtual image with the real, lighted surroundings is the main drawback of see-through devices. This is supported by the fact that the enhanced images are neither too distracting or too dominant (Figure 4.6, [29]).

Augmented reality (AR) uses see-through optics to connect the virtual with the physical environment, yet this design results in significant energy loss from the display lights. Dimming the lights in the room and using a dark background for the holograms is one way to improve the visibility of

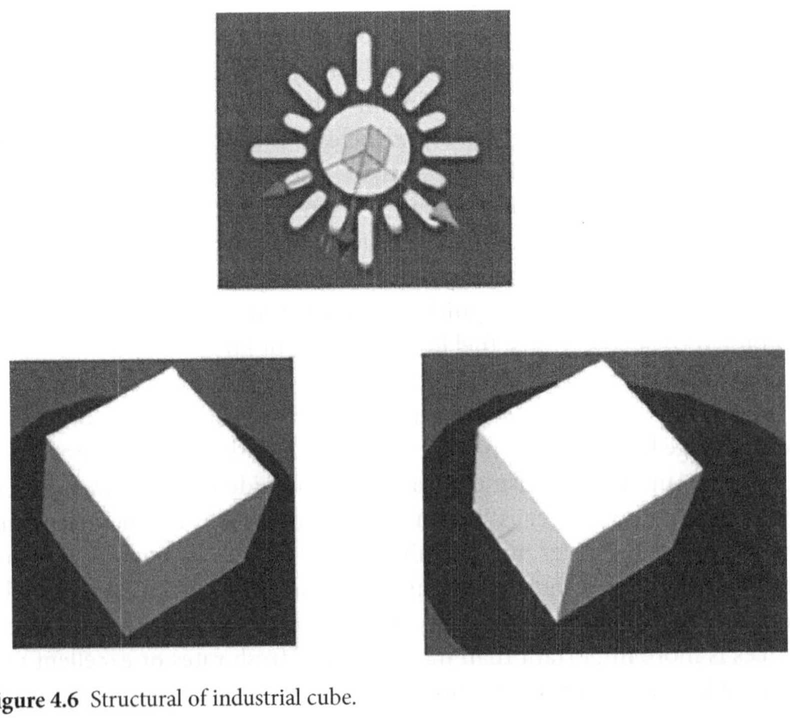

**Figure 4.6** Structural of industrial cube.

3D objects, but this eliminates the "see-through" feature. The see-through display of the HoloLens, for instance, can only increase the brightness of the user's field of view, thus any virtual objects seen through it would seem muted. There is no additional layer that might potentially black out the display in certain locations. The display is so bright that virtual images almost completely obscure the environment in dim demo rooms, yet they become almost invisible when seen next to a bright background (such as a table lamp lying nearby). Optical see-through technologies, as specified in the instructions for augmenting the user's vision with light, have trouble displaying fully black 3D objects (RGB = [0,0,0]) due to the fact that black is used as an alpha channel (the layer used for transparency). In the Unity Editor, you may change the transparent hue of HoloLens by choosing "Clear Flag" and then "Solid Black" from the main camera's pop-up menu. The Meta2 framework [30] has similar advantages and disadvantages.

The brightness problem, which also affects visual quality, is the main difference between these and video see-through HMDs. In particular, holograms seem to "glow" and lose detail when seen via optically transparent HMDs. Yet because to their OLED displays and eyepieces, devices like the HTC Vive and Oculus Rift don't have this problem. In addition, VST technology solved the occlusion problem of HMDs before OST HMDs did. Hence, the OST technology precludes the usage of eyepieces, which are necessary for peripheral coverage. It is possible to turn off pixels with VST devices utilizing computer vision libraries, allowing one to either substitute real things with improved ones or represent them as completely black (such as OpenCV). In contrast to the cutting-edge Meta2 projection system, which employs a half-silver convex mirror to project images from an LCD screen straight into the user's eye, the older method relies on the tried-and-true method of using a mirror. Meta2's projection technology provides a wider viewing area. Nevertheless, if the distance here between eye and the white beveled is too high, the distortion factor of the lenses will diminish the see-through capability. The increased computational power of video see-through HMDs is another key differentiator between optically and video see-through HMDs. Embedded computing is only available on the Microsoft HoloLens and a few Vuzix models with Intel Atom Processors. In order to use software, both the HTC Vive and the Oculus Rift need a personal computer, and the headsets must be physically connected to the computer through a cable. The tracking system also requires additional hardware. In a manufacturing scenario, they are not ergonomically viable. Microsoft's software, on the other hand, offers more portability for use in a mobile setting since it does not need an active internet connection. Nevertheless, its spatial surveillance technology, which connects each of the rooms to a Wi-Fi access point, is useless without a stable and fast wireless network. HoloLens retrieves these meshes from storage when it detects a connected Wi-Fi network. The abovementioned difficulties make Microsoft HoloLens not an easy choice for the industrial scenario [31].

### 4.3.3 Limitations of CAD Model

Because of the various obstacles that must be overcome, it would be a stretch to argue that Augmented Reality, as opposed to Augmented reality is a mature, implementable technologies with many potential uses in industry. This is owing to the fact that users may experience annoyance and lost time as a result of the large variety of software applications and frameworks that aren't always connected with one another. The time spent preparing for a single animation is substantial. For a practical industrial application, the authors recommend beginning with the production of a 3D parametric CAD model, then exporting that model as a mesh, simplifying the geometry tool, and last importing that model into Unity to begin scripting Augmented Reality for the final scenario. This sequence of steps obviously requires not only a significant investment of time but also an in-depth familiarity with every resource at your disposal. Real-time rendering polygon constraints also mean that end users have to settle for less detailed models than they'd like because of this time sink, the technology isn't yet ready for usage in maintenance manuals. It's just too time-consuming to create animations for each and every potential duty that operators could face. Particularly relevant here is PTC Creodedications to unifying 3D model production with augmented reality visualization in Figure 4.7. Vuforia's integration with Parametric PTC CAD and ThingWorx Studio's part in establishing the necessary environment make this feasible. Notwithstanding the limits of the environmental scene compared to Unity Editor, this effort may be viewed as a huge step towards the use of AR in true industrial applications, not only in games or computer graphics. Notwithstanding such facts, the most prevalent use of AR at the present time is in training, not management or the implementation of virtual guides. So, with the use of AR, it is no longer necessary to schedule training sessions around the teacher's schedule [32].

### 4.3.4 Augmented Reality in Manufacturing Sector

Many usability difficulties, in addition to software and hardware constraints, must be considered when using augmented reality in a manufacturing setting. Current augmented reality applications frequently need

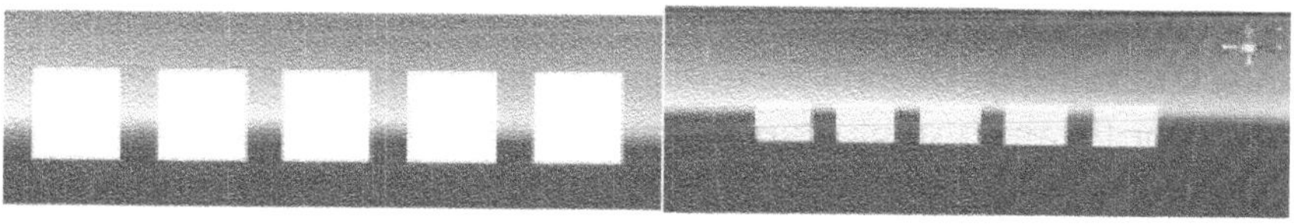

**Figure 4.7** A solid black cube framed.

set-up activities prior to beginning session of labor; acceptable in a laboratory context but inappropriate for an office worker who is not an expert in computer animation. In addition, industries cannot afford to lose time on laborious and drawn-out set up processes. Moreover, current augmented reality applications' inability to easily handle collision/occlusion is a problem in the business world. Occlusion and collision effects are essential for providing a sense of depth, realism, and enabling maintenance operations that require the movement and orientation of 3D parts in space. Light levels may be a major issue for augmented reality equipment in factories with both dark and bright areas, where the same zone could move from one lighting level to another just by switching on or off a light or opening a door. It's worth noting that the current crop of touch-based command/input windows isn't reliable enough for use in manufacturing [33, 34]. Users in an industrial context simply do not have the luxury of time to experiment with and master various input methods. While they are not yet ubiquitous in augmented reality applications, finger select and drag functions may help increase usability. No wearable augmented reality glasses suitable for tasks requiring significant computing effort are presently on the market. There are now two types of glasses on the market: ergonomic ones with low computational capacity and subpar visuals, and high-performance Head Mounted Displays that may be bulky and uncomfortable to wear. Actually, there aren't any high-resolution, analytically capable glasses available right now that also have good ergonomics. These assertions stem from "first-person" testing conducted by the authors themselves in a business context. These are provided to give the reader a better grasp of the concepts discussed in the text. Figure 4.8a depicts a user in a simulated workshop environment while wearing augmented reality drilling goggles. As can be seen in Figure 4.8b, the user is then given with the photos showed.

Nevertheless, the 0.05 mm accuracy needed for drilling that is common in mechanical production is very difficult to attain, and even the 1 mm precision that is now anticipated (with marker and marker-less technology) requires a time-consuming registration procedure. Also, the brightness of the scene experienced while wearing the glasses is affected by external factors; for example, AR software does not make allowances for the opening of a door (or the turning on of a fluorescent lamp). Lastly, even a simple augmented reality animation requires a lot of time to prepare utilizing a dedicated augmented reality software application. The scenario was made by importing the CAD model into Vuforia, aligning it with the marker (see Figure 4.9), adding a symbol, and then loading the animations into the Hololens glasses.

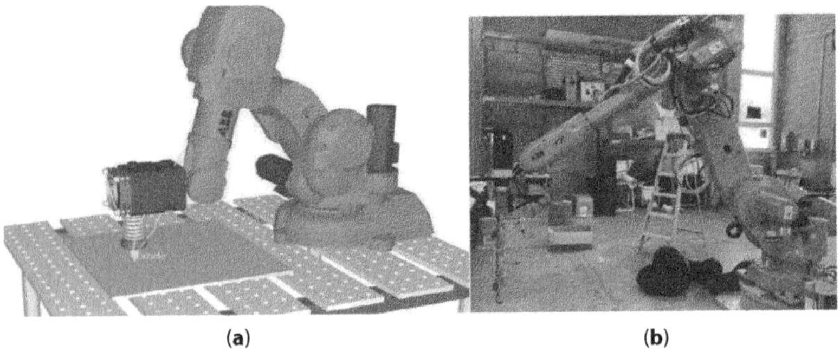

**Figure 4.8** 3D object positioned in real world.

**Figure 4.9** AR scene

The average time needed by a professional operator to construct the animation was determined to be 2 hours after a series of tests with many users; a marker-less technique is also possible. For the latter, the environment serves as a marker in and of itself, making sterile environments flooring and walls that are completely flat and featureless, as well as a bland color scheme, are not ideal for computer vision and posing. The complex shape and easily recognizable parts with edges and multi-oriented surfaces of most machines make this a rare problem in a manufacturing.

## 4.4 Requirement of Augmented Reality

### 4.4.1 Capability of AR

Once the whole Reality-Virtuality continuum is reached, the authors think it will be possible to effectively and efficiently leverage AR characteristics in an industrial setting. This concept, first presented by Paul Milgram, includes allowing the user to decide how much of the virtual world they want to interact with. In AR commercial applications, the ability to switch between the both the real and virtual worlds is vital, as is the operator's ability to concentrate on either the virtual or real worlds, according to the demands of the task at hand. The border between the real and the virtual may be blurred with the use of many technologies that should be made available to users. These technologies include shadows, correctly integrated occlusion, generated visual representations, and highly competent human-machine interfaces. As will be made clearer below, this shift in technology has consequences for the computer systems and programs needed to operate augmented reality.

### 4.4.2 Computational Hardware Capabilities

The computational capabilities of glasses need to be upgraded in terms of hardware. Embedded computational power is currently only found in the Microsoft HoloLens and a few Vuzix versions. Even if other gadgets are convenient, they lack processing power and need a connection to other gear. Unfortunately, the HoloLens and Vuzix Glasses' installed computing capacity is limited, and the gadgets fail when they need to analyze complex geometry and are not sufficiently simplified, in addition to being fairly clunky and unpleasant when used in an industrial setting. Freedom of movement in the physical world is the primary advantage of having one's own processing capacity (i.e. on-field applications). Technologies like Meta2, which use the processing capacity of a distant server, provide improved performance at the cost of restricted user movement. Even with a backpack, the recommended hardware for using tethered devices is quite demanding. For instance, the latest generation of processors and specialized high-end graphics cards, which are required for operating Meta2, are not often included in a laptop of moderate size and weight. However, other devices like as the HTC Vive and the Oculus Rift could not be used in an outside setting, even in a backpack configuration, since the monitoring system needed external sensors. Separating the HMD from the semiconductors used in computations is another way that research has suggested to preserve mobility. One such solution is proposed by Magic Leap [35], which utilizes a tiny, wearable computer to do computations, completely apart from the display. As soon as advances in electrical technology allow for the production of small, powerful processors with dedicated graphics processing units (GPUs), the processor will be included into the frame of the glasses themselves, mitigating the need to choose between portability and processing capabilities. The problems created by ambient lighting should be resolved in the next generations of optically transparent HMDs, paving the way for their use in outdoor situations. When the problem of ambient light interference is fixed, physical objects may be successfully hidden, and their digital twins can step in to fill the void. Improving the real-virtual data exchange requires the integration of the technologies needed to scan physical things and automatically transfer them into the virtual world [36] foresee that augmented reality will be employed to increase output if and when these capabilities become available. Nevertheless, a higher degree of precision in AR scene identification is required when assessing the tolerance range for manufacturing dimensional tolerances. People's in-use interactions with physical objects are also altered by the real-virtual data stream. Industry 4.0 primarily depends on the concept of the digital system, which involves engaging with the virtual model by operating on the entire model (and vice versa) and immediately using the collected data to develop superior engineering decisions and designs based on real-world data. The resolution, frame rate, and refresh rate upgrades are needed to reduce or eliminate the wake effect during viewpoint changes. Improved visibility of virtual objects, especially their edges in 3D, is made possible by higher display resolution. Refresh rates of 30 frames per second (fps), which are now available in many hardware devices, are sufficient since they are faster than those involved in an activity observed by human eyes. The appropriate resolution settings for a display are context-specific. Ferwerda [37] categorizes the realism of computer graphics into three levels: physical realism (using the same sensory information as the scene), photo realism, and operational realism (identical visual information as the scene). HoloLens and related devices should be used in maintenance, where a high degree of realism is required without compromising on functionality. To achieve such photorealism, higher resolutions are necessary. One possible solution to the problem of correctly presenting entirely black objects is to employ eyeglasses or glasses that hide the black zones, as well as new technologies like LCD electrochromic films.

### 4.4.3 Symbol-Based Tracking Software

There are numerous well-documented panels in factories, making symbol-based tracking a viable option for industrial applications; comparable technology is already in use in Laser-Guided Vehicles (LGV). Another technology that may be utilized for use in the absence of a human operator in industrial settings requiring autonomy is SLAM (simultaneous localization and mapping). The primary goal of all these apps is software simplicity, which necessitates adaptable integration among widely used popular programs. To bridge this gap, we propose the introduction of plugins that allow AR libraries to communicate with commercial CAD without requiring users to manually convert models and create scenes in several applications. PTC has previously introduced their Vuforia Object Tracking [38] solution for this problem.

In the wake of its purchase of Vuforia, PTC established a bridge between its augmented reality (AR) and computer-aided design (CAD) software, allowing for streamlined AR scene preparation. Because of object tracking, CAD models may be automatically matched with real-world objects without any human intervention. With this connection, augmented reality apps may be used in the most intuitive manner imaginable. Besides this, it's worth noting that augmented reality (AR) is superior than virtual reality (VR) for usage in the workplace since it's less intrusive and facilitates a more natural integration with the actual world. Although Virtual Reality (VR) has its place in training, it is now hard to imagine Real-world industrial uses for virtual reality, where handling real objects is essential. One way in which AR technology excels over VR is in the superior picture and hologram quality it makes possible. Nevertheless, picture quality should be a top priority. When AR was originally being created, solving the registration issue was a top priority; now, because to advances in computer power and tracking technologies, AR developers can now focus on making their AR experiences seem more natural by improving the realism of the projected items. With their idea, VTT [39] has made headway in solving real-time illumination. It's not a perfect answer, but it's a step in the right direction towards a more precise model. Real and virtual items should be totally convertible, and new software packages should facilitate reverse engineering of things captured in camera frames. Real "digital twin" capabilities may be attained by improving the registration problem [40] and rendering methods. One of the most important aspects of merging the physical and digital worlds is a thorough comprehension of the surroundings (through scene reverse engineering and scanning). Finally, AR requires devices that are lightweight, so that users may work in comfort. Since it requires less processing power, augmented reality may be easily transported, opening up new possibilities in the manufacturing sector.

## 4.5 Conclusions

The integration of AI-based AR/VR models in the biomedical field, particularly in the context of sustainable Industry 4.0 for bioprinting services, offers significant benefits and transformative potential. This abstract has explored the various applications and advantages of this integration, emphasizing the advancements it brings to bioprinting processes and its contribution to the overall sustainability of the industry. By leveraging AI algorithms, AR/VR models can analyze vast amounts of data and generate optimized bioprinting patterns and strategies. This leads to improved accuracy, productivity, and efficiency in the bioprinting workflow. Real-time monitoring and adaptive control capabilities enable immediate adjustments and reduce errors, enhancing the quality of the printed tissues and organs. Moreover, the immersive visualization and interaction provided by AR/VR technologies facilitate better design and validation processes. Researchers and clinicians can assess the functionality and compatibility of bioprinted structures before implementation, leading to more successful outcomes and reducing the need for costly and time-consuming iterations. In the broader context of sustainable Industry 4.0, the integration of AI-based AR/VR models contributes to the overall sustainability of bioprinting services. By optimizing the printing process, minimizing material waste, and enabling continuous learning and improvement, these technologies reduce the environmental impact of bioprinting. Additionally, the personalized and functional tissues and organs produced through this approach have the potential to revolutionize regenerative medicine and improve health care outcomes. However, it is important to acknowledge the challenges and considerations associated with implementing AI-based AR/VR models in the biomedical field. Ethical considerations, data privacy, and regulatory frameworks must be carefully addressed to ensure the responsible and secure use of these technologies. In conclusion, AI-based AR/VR models hold great promise for the sustainable advancement of bioprinting services in the biomedical field. Their integration within the framework of Industry 4.0 enables improved accuracy, productivity, and efficiency, while also contributing to environmental sustainability. Continued research and development in this area will likely lead to groundbreaking applications, ultimately transforming the landscape of regenerative medicine and health care as a whole.

## References


1. Qadri, Y.A., Nauman, A., Bin Zikria, Y., Vasilakos, A.V., Kim, S.W., The future of healthcare internet of things: A survey of emerging technologies. *IEEE Commun. Surv. Tutor.*, 22, 1121–1167, 2020.
2. Qian, J., Feng, S., Tao, T., Hu, Y., Li, Y., Chen, Q., Zuo, C., Deep-learning-enabled geometric constraints and phase unwrapping for single-shot absolute 3D shape measurement. *APL Photonics*, 5, 046105, 2020.
3. Gattullo, M., Evangelista, A., Uva, A.E., Fiorentino, M., Gabbard, J., What, how, and why are visual assets used in industrial augmented reality? A systematic review and classification in maintenance, assembly, and training (from 1997 to 2019). *IEEE Trans. Vis. Comput. Graph.*, 28, 2, 1143–1456, 2020.
4. Kaswan, K.S., Dhatterwal, J.S., Kumar, K., Blockchain of internet of things-based earthquake alarming system in smart cities, in: *Integration and Implementation of the Internet of Things through Cloud Computing*, pp. 272–287, IGI Global, Hershey, Pennsylvania, USA, 2021.
5. Au-Yong-Oliveira, M., Pesqueira, A., Sousa, M.J., Dal Mas, F., Soliman, M., The potential of Big Data research in healthcare for medical doctors' learning. *J. Med. Syst.*, 45, 1, 1–14, 2021.
6. Neges, M., Adwernat, S., Abramovici, M., Augmented Virtuality for maintenance training simulation under various stress conditions. *Procedia Manuf.*, 19, 171–178, 2018.
7. Pang, Z., Yang, G., Khedri, R., Zhang, Y.T., Introduction to the special section: Convergence of automation technology, biomedical engineering, and health informatics toward healthcare 4.0. *IEEE Rev. Biomed. Eng.*, 11, 249–259, 2018.
8. Dhatterwal, J.S., Kaswan, K.S., Jaglan, V., Vij, A., Machine learning and deep learning algorithms for IoD, in: *The Internet of Drones*, pp. 237–292, Apple Academic Press, New Jersey and Canada, 2022.
9. Amin, D. and Govilkar, S., Comparative study of augmented reality Sdk's. *Int. J. Comput. Sci. Appl.*, 5, 11–26, 2015.

10. Celdrán, A.H., Pérez, M.G., Clemente, F.J.G., Pérez, G.M., Sustainable securing of medical cyber-physical systems for the healthcare of the future. *Sustain. Comput. Inform. Syst.*, 19, 138–146, 2018.
11. Kounavis, C.D., Kasimati, A.E., Zamani, E.D., Enhancing the tourism experience through mobile augmented reality: Challenges and prospects. *Int. J. Eng. Bus. Manage.*, 4, Digital and Mobile Economy, 10, 2012. www.intechopen.com
12. Klimova, A., Bilyatdinova, A., Karsakov, A., Existing teaching practices in augmented reality. *Procedia Comput. Sci., 7th International Young Scientist Conference on Computational Science*, 136, 5–15, 2018.
13. Gattullo, M., Dammacco, L., Ruospo, F., Evangelista, A., Fiorentino, M., Schmitt, J., Uva, A.E., Design preferences on industrial augmented reality: A survey with potential technical writers, in: *Proceedings of the 2020 IEEE International Symposium on Mixed and Augmented Reality, ISMAR-Adjunct 2020*, Recife, Brazil, pp. 172–177, November 9–13, 2020.
14. Kaswan, K.S., Dhatterwal, J.S., Baliyan, A., Jain, V., Special sensors for autonomous navigation systems in crops investigation system, in: *Virtual and Augmented Reality for Automobile Industry: Innovation Vision and Applications*, pp. 65–86, Springer International Publishing, Cham, 2022.
15. Zhu, E. *et al.*, Augmented reality in healthcare education: An integrative review. *Peer J.*, 2, e469, 2014.
16. Goh, E.S., Sunar, M.S., Ismail, A.W., 3D object manipulation techniques in handheld mobile augmented reality interface: A review. *IEEE Access*, 7, 40581–40601, 2019.
17. Beattie, N., Horan, B., McKenzie, S., Taking the LEAP with the oculus HMD and CAD—Plucking at thin air? *Procedia Technol.*, 20, 149–154, 2015.
18. Agnusdei, G.P., Elia, V., Gnoni, M.G., Is digital twin technology supporting safety management? A bibliometric and systematic review. *Appl. Sci.*, 11, 2767, 2021.
19. Kaswan, K.S., Dhatterwal, J.S., Preeti, K., Intelligent agent based case base reasoning system build knowledge representation in COVID-19 analysis of recovery infectious patients, in: *Application of AI in COVID 19*, 2020.
20. De Marchi, L., Ceruti, A., Marzani, A., Liverani, A., Augmented reality to support on-field post-impact maintenance operations on thin structures. *J. Sens.*, 2013, 1–10, 2013.
21. Mourtzis, D., Siatras, V., Angelopoulos, J., Real-time remote maintenance support based on augmented reality (AR). *Appl. Sci.*, 10, 1855, 2020.
22. Galati, M., Snis, A., Iuliano, L., Powder bed properties modelling and 3D thermo-mechanical simulation of the 766 additive manufacturing electron beam melting process. *Addit. Manuf.*, 30, 100897, 2019.
23. Baron, L. and Braune, A., Case study on applying augmented reality for process supervision in industrial use cases, in: *Proceedings of the IEEE International Conference on Emerging Technologies and Factory Automation, ETFA*, Berlin, Germany, September 6–9, 2016.

24. Di Donato, M., Fiorentino, M., Uva, A.E., Gattullo, M., Monno, G., Text legibility for projected Augmented Reality on industrial workbenches. *Comput. Ind.*, 70, 70–78, 2015.
25. Maly, I., Sedlacek, D., Leitao, P., Augmented reality experiments with industrial robot in Industry 4.0 environment, in: *Proceedings of the IEEE International Conference on Industrial Informatics (INDIN)*, Poitiers, France, pp. 176–181, July 18–21, 2016.
26. Dhatterwal, J.S., Kaswan, K.S., Pandey, A., Implementation and deployment of 5G-drone setups, in: *The Internet of Drones*, pp. 47–64, Apple Academic Press, New Jersey and Canada, 2022.
27. Hanson, R., Falkenström, W., Miettinen, M., Augmented reality as a means of conveying picking information in kit preparation for mixed-model assembly. *Comput. Ind. Eng.*, 113, 570–575, 2017.
28. Reif, R., Günthner, W.A., Schwerdtfeger, B., Klinker, G., Evaluation of an augmented reality supported picking system under practical conditions. *Comput. Graph. Forum*, 29, 2–12, 2010.
29. Schwerdtfeger, B., Reif, R., Günthner, W.A., Klinker, G., Pick-by-vision: There is something to pick at the end of the augmented tunnel. *Virtual Real.*, 15, 213–223, 2011.
30. Xu, M., Li, T., Wang, Z., Deng, X., Yang, R., Guan, Z., Reducing complexity of HEVC: A deep learning approach. *IEEE Trans. Image Process.*, 27, 5044–5059, 2018.
31. Pratt, P., Ives, M., Lawton, G., Simmons, J., Radev, N., Spyropoulou, L., Amiras, D., Through the HoloLens™ looking glass: Augmented reality for extremity reconstruction surgery using 3D vascular models with perforating vessels. *Eur. Radiol. Exp.*, 2, 1–7, 2018.
32. Tepper, O.M., Rudy, H.L., Lefkowitz, A., Weimer, K.A., Marks, S.M., Stern, C.S., Garfein, E.S., Mixed reality with hololens: Where virtual reality meets augmented reality in the operating room. *Plast. Reconstr. Surg.*, 140, 1066–1070, 2017.
33. Wang, Z., Bai, X., Zhang, S., Billinghurst, M., He, W., Wang, Y., Han, D., Chen, G., Li, J., The role of user-centered AR instruction in improving novice spatial cognition in a high-precision procedural task. *Adv. Eng. Inform.*, 47, 101250, 2021.
34. Lee, G.A., Ahn, S., Hoff, W., Billinghurst, M., Enhancing first-person view task instruction videos with augmented reality cues, in: *Proceedings of the 2020 IEEE International Symposium on Mixed and Augmented Reality, ISMAR 2020*, Porto de Galinhas, Brazil, pp. 498–508, November 9–13, 2020.
35. Wang, B., Zhang, B.F., Liu, X.W., Zou, F.C., Novel infrared image enhancement optimization algorithm 793 combined with DFOCS. *Optik*, 224, 165476, 2020.
36. Popov, V., Muller-Kamskii, G., Kovalevsky, A., Dzhenzhera, G., Strokin, E., Kolomiets, A., Ramon, J., Design and 865 3D-printing of titanium bone

implants: Brief review of approach and clinical cases. *Biomed. Eng. Lett.*, 8, 337–344, 2018.
37. Ferwerda, J.A., Three varieties of realism in computer graphics. *Human Vision and Electronic Imaging VIII*, vol. 5007, p. 290, 2003.
38. Richler, J., The translational potential of virtual reality. *Nat. Rev. Psychol.*, 1, 12, 2022.
39. Aittala, M., Inverse lighting and photorealistic rendering for augmented reality. *Visual Comput.*, 26, 669–678, 2010.
40. Datta, S.P.A., Emergence of digital twins, in: *Internet of Things Collaborative Research Initiative (ICRI) at MIT Auto-ID Labs*, 2017.

# 5

# Computational Intelligence–Based Image Classification for 3D Printing: Issues and Challenges

**Jagjit Singh Dhatterwal[1*], Kuldeep Singh Kaswan[2], B. Tirapathi Reddy[3] and Gangadevi E.[4]**

[1]*Department of Artificial Intelligence & Data Science, Koneru Lakshmaiah Education Foundation, Vaddeswaram, Andhra Pradesh, India*
[2]*School of Computing Science and Engineering, Galgotias University, Greater Noida, India*
[3]*Department of IT, P.V.P Siddhartha Institute of Technology, Vijayawada, Andhra Pradesh, India*
[4]*Department of Computer Science, Loyola College, Chennai, India*

***Abstract***
Three-dimensional (3D) printing has emerged as a promising technology with applications across various domains, including manufacturing, healthcare, and consumer products. The quality and accuracy of 3D printed objects heavily rely on the design and preparation of the digital models. Image classification techniques, powered by computational intelligence (CI) methods, have the potential to enhance the accuracy and efficiency of 3D printing by automating the classification of digital images. However, several challenges and issues need to be addressed to successfully implement CI-based image classification in the context of 3D printing. This chapter presents an abstract that outlines the primary issues and research challenges associated with implementing computational intelligence-based image classification for 3D printing. First, the paper discusses the complexities involved in the classification of digital images in the context of 3D printing, including the diversity of object shapes, sizes, and materials. Next, it highlights the need for a robust and accurate feature extraction process that can capture relevant information from the images for classification. Furthermore, the paper addresses the challenge of training data availability and quality. Due to the scarcity of labeled datasets specific to 3D

**Corresponding author*: jagjits247@gmail.com

E. Gangadevi, M. Lawanya Shri, Rajesh Kumar Dhanaraj and Balamurugan Balusamy (eds.) Computational Intelligence in Bioprinting, (79–92) © 2024 Scrivener Publishing LLC

printing, it becomes necessary to explore transfer learning techniques and develop methods to generate synthetic training data. Additionally, the research community needs to focus on addressing the issue of class imbalance, as certain object classes might be significantly underrepresented in the training data. Moreover, the paper discusses the computational requirements and efficiency concerns associated with CI-based image classification for 3D printing. Developing computationally efficient algorithms and optimizing the training and inference processes becomes crucial to ensure real-time performance and feasibility in resource-constrained environments. Finally, the chapter emphasizes the importance of interpretability and explainability in CI-based image classification for 3D printing. Understanding the decision-making process of the classification algorithms is vital for gaining user trust and enabling human-machine collaboration in critical applications.

***Keywords:*** Machine learning, 3D printing, artificial intelligence, additive manufacturing

## 5.1 Introduction

Artificial intelligence (AI) has recently surfaced as a potential solution to the issues of the future, and it is already being implemented into several state-of-the-art equipment. The entire potential of AI can be realized in 3D printing. This is unquestionably state-of-the-art equipment. Technology, such as artificial intelligence, is currently being used to improve 3D printers. The use of AI into the 3D printing process will greatly accelerate the development of the technology [1].

One sort of intelligence shown by machines is artificial intelligence, also called machine intelligence. This kind of machine is capable of doing difficult jobs because it learns and thinks like a person by collecting data autonomously, analyzing it, and drawing conclusions. Intelligent devices simulate human intellect. The AI automation process may be performed independently in a variety of methods [2].

Although operating a 3D printer is complicated in and of itself, adding AI has the potential to significantly boost the efficiency of the technology. In recent years, "3D printing" has evolved to refer to the method of fabricating a three-dimensional object by adding successive layers from a digital model. Many problems, such as long wait times for traditional prostheses and, in rare situations, a lack of availability, may be resolved by incorporating 3D printing into upper-limb prostheses [3].

John Wesley Hyatt discovered cellulose in 1868 while searching for a new material for billiard balls, launching the plastic materials industry. In 1950, a firm named Akerwerk in Germany developed the very first injection molding machine.

The introduction of plastics into manufacturing caused a sea change. Because to advancements in technology, plastics are now malleable polymers that may be used in a myriad of applications. The future of manufacturing will be drastically different as a result of the integration of AI into the design of 3D printing facilities. The London company AI Build has developed a 3D printing platform that is driven by AI and may be used for automated manufacturing [4].

The smart extruder is built in and can identify issues as they arise. It can think and act on its own, too. It is a massive 3D printer powered by AI software and factory robots. The use of robotic arms, particularly ones that can independently identify and print working components, represents a significant step forward in manufacturing. AI Build is creating self-correcting robots for construction by teaching them to observe, plan, and fix their errors [5].

The rest of the paper is broken out as follows: The second part sets the stage for the arguments made in the third part. In the third part, we survey the big picture of the various artificial intelligence approaches employed in 3DP. A cloud service platform, service assessment, and detection techniques are all a part of this. Other features include assessing printability and accelerating slicing and designing nozzle paths. In Section 4.4, we explore the results and their relevance [6].

## 5.2 Brief Concepts

Additive manufacturing (AM), artificial intelligence (AI), and machine learning are all introduced here.

### 5.2.1 3D Printing Tools

The field of additive manufacturing has quickly risen to become one of the most exciting and respected in the modern technological landscape. In 1984, the stereolithographic printing technique was created, which set the door for the growth of 3D printing tools. Additive manufacturing (AM) allows for the production of objects with complex geometry that would otherwise be impossible to produce [7].

Rapid prototyping and digital manufacturing are two ways in which AM propels the development of science. In engineering applications, a broad variety of polymer, metal, and biomaterials are created and fabricated to create prototypes and final products with unique shapes, multifunctional compositions, reliability, and high quality. The development of AM technologies will have far-reaching effects on the future [8].

Reduced inventory requirements and streamlined distribution channels are the results of recent technology developments that make it possible to print nearly anything at home or in the workplace.

As we now live in a digital era of manufacturing, ideas may be translated to 3D models and sent to a 3D printer, which can start to work and finish without any human interference. Consumers are only interested in products that are directly related to the production of food [9].

### 5.2.2 Artificial Intelligence–Based Digital Marketing

The creation of AI and related software has the greatest promise in terms of both job creation and consumer interest. There has to be a worldwide transition from the old market to the digital market [10].

The additive manufacturing marketplaces of the world's countries will be united as they work together to develop and export 3D printing technology to other regions of the globe. Several of the most developed nations in Europe and the United States have been early adopters of digital manufacturing. AM is utilized in the production of the vast majority of airplane parts [11].

### 5.2.3 Automated Machine Learning Prediction System

The machine learning subfield has expanded rapidly as a result of increased data accessibility. The field of artificial intelligence known as machine learning focuses on the creation of data-driven automated learning and prediction systems [12]. Seeing patterns in large datasets and extrapolating conclusions or valuable information from them is a common use of algorithms. The positive consequences of machine learning may be seen in many different areas of society, including big data, the IoT, the web, psychology, medicine, and many more.

## 5.3 Role of Artificial Intelligence in Industry 4.0

In recent years, both machine learning and additive manufacturing (AM) have received increased focus from academics, industry, and the general public. Industry 4.0, also referred to as intelligent fabricating or automation fabrication, is a movement led by a variety of industries with the goal of ushering in the next generation of production. Here is where Amisa woman's creative and innovative technology comes Current AM premanufacturing processes use ML methods to solve problems. Production by means of observation and trial and error [13].

This session of papers examines recent machine learning (ML) research on AM. To check for malicious in fill problems, high-resolution cameras were put at the front, top, and left of the 3D desktop in the level printers. The preview of the simulation was used to capture the images in order. In order to examine malicious 3D printing problems, photos of experimental items were analyzed, both with and without defects present in the infills. Classification and decision tree J48, both forms of machine learning (ML), were used to analyze the photographic data [14].

According to the findings, Naive Bayes achieves an accuracy of 85.26%. Tree-based methods for classification and decision making, with a J48 accuracy of 95.51%. Promotion and employment of 3DP are still restricted in contrast to typical production techniques because to the geometric characteristic, consumption of time, and particular material demand of 3DP [15].

### 5.3.1 3D Printing Process

To ease the manufacturing process and ensure that the 3D model can be printed optimally, the printability checker (PC) method was developed to determine whether a product is suitable for 3D printing or other types of production.

### 5.3.2 Enhancement in Machine Learning

There is still work to be done before the real implementation can be deemed entirely functional and dependable, particularly when dealing with a large range of features and data, but the experiments show that the PC technique is effective in evaluating the printed model. Specifically for the printability investigation, Lu recommended using the ML technique for automating the modification of pertinent laws and parameters [16].

The support vector machine (SVM) is an alternative method that may be used to train the estimation model for printability predictions. It is possible to arrive at the optimum decision function by repeatedly classifying with the same set of parameters. Furthermore, experimental results showed that the technique successfully reduced the time needed for feature extraction in 3D models without compromising accuracy [17].

### 5.3.3 Genetics-Based Machine Learning

Assessment of the printability or complexity of a company engaged requires consideration of several elements, such as time, cost, raw material, model size, and geometry. Testing and identifying multi-indicators, as well as what effect percentage each indication has, may be made easier with the use of genetic algorithms (GAs) and the genetics-based ML technique, which can maximize the impact proportions and the least complexity value.

The reason for this is that GA functions best in large environments or with data that can be encoded as a binary sequence. This probability search method of building optimization techniques requires fewer starting assumptions. Moreover, cutting-edge developments in 3D printing are the foundation of the PC approach. If 3D printing is to become widely used, further research is required to optimize the process control to lower the complexities threshold in a multi-indicator setting. Improved printing technology and a printability tester have led to the printableization of a growing variety of products [18].

### 5.3.4 Slicing Technique in 3D Model

In order to transform the sliced 3D model into a collection of slicing planes in z coordinates, the layer information must be retrieved from the triangular mesh using a slicing technique. Three kernel modules—ray-triangle intersection (TRI), trunk sorting (TS), and layer extraction—comprise the slicing technique presented by Wang *et al* (LE). Using TRI, the slicing algorithm can determine where the vertical rays from the centers of the 2D image's pixels cross with the triangles in the STL mesh. Finding the plane's intersection uses similar techniques to those described in [19].

The goal of TS is to provide a rational order for the boot intersections. Using the layer height and the position of the point, LE calculates the binary value of each pixel and then constructs the layer images to be printed. While the slicing method lays the framework for additional layer-based content manufacturing, it has a number of limitations in terms of processing expense and the difficulty of parallel implementation. Image enhancement, production rules, logic automation, data preprocessing, data mining, etc. all have significant computational requirements, and with the advent of big data, virtual memory has appeared as a possible technique of achieving these demands [20].

To hasten this prefabrication process, proposed two GPU techniques: pixel-wise parallel slicing (PPS) and totally parallel slicing (FPS). The eventual objective of PPS is parallelization of the slicing technique this facilitates the GPU's ability to support the threads and allocate them to each pixel ray. Each ray will then be handled by its own separate thread. The threads' use of GPU shared memory for tracking intersections and sorting results also contributes to the time savings. FPS extends the parallelism of RTI, TS, and LE, especially when it comes to solving large-scale slicing problems. In contrast to PPS, FPS makes use of a multithreaded method to simultaneously operate three independent modules [21].

### 5.3.5 Printing Path Trajectory

In addition to directing the nozzles to produce the better visualization, efficient printing trajectories may cut down on computation and printing time. Most of the time, a printing nozzle will be transitioning between print segments and other types of segments. Thus, it is essential to locate the most expedient path, whether in terms of time or distance.

Fok *et al.* introduced a 3DP route optimizer that uses a relaxation strategy inspired by the Traveling Salesman Problem to determine the time required to traverse a nozzle (TSP). The path expertise and knowledge may be recast as a traveling salesman problem by equating each newspaper's circulation area with a city and asking for the optimal route across the country (TSP). TSP improves the quality of the edges between the layers of print (interpartition level) and nonprint (intrapartition level).

TSP can figure out the best time to go between interpartitions and set the wheels in motion for the journey. The Christofides approach is then used to compute the optimum time spent at this location. A transition segment connects nearest endpoints in adjacent print areas at the intrapartition level [21].

### 5.3.6 Improvement in Computational Simulation

The system in may simplify and accelerate the printing process, as proved by simulation software; nonetheless, it has a few drawbacks that may affect print quality and the optimal solution. To begin, if one uses distance as a criterion, then the Christofides approach is the sole way to prove that the nozzle's traversal distance or time is optimal. Nevertheless, if we ignore the time it takes to retract, it loses its claim to being the fastest option.

Retraction is extensively explored since excess filament always escapes from a traditional printer. It takes time for the filament to retract while the nozzle travels between segments. As this retraction time increases linearly with the number of segments, it must be confirmed via modeling and experimentation before being utilized as a constant.

Second, researchers developed a simplified method of combining neighboring print sections based on a consolidating threshold value, which speeds up both the computing and printing operations. But, there is a problem with the threshold control of this approach. This is because a large threshold may create geometry changes at the print area's edge, which can affect the smoothness of the transition between adjacent interpartitions or the precision of the final 3D object.

Future work should focus on how and to what extent threshold control can achieve the optimum printing speed without sacrificing accuracy. Accelerating parallel slicing and optimizing pathways has been proven in several research to potentially provide excellent productivity. Yet, there have been very few investigations on how to efficiently merge printability testing, slicing, and route planning. To further accelerate the prefabrication, the layer-based printing technique suggests using parallel processing to both convert the 3d representation into a layer representation and autonomously construct a suitable path.

### 5.3.7 Improving Service-Oriented Architecture

The computer paradigm known as service-oriented architecture (SOA) is crucial to cloud manufacturing because it provides the means to swiftly and efficiently respond to the requirements of customers. Technically feasible SOA may intelligently realize 3DP's tremendous flexibility, integration, and customizability. Scholars have investigated the problem of correlation across virtual services with varying prerequisites and constraints.

For instance, Li *et al.* discussed the results of cloud manufacturing, but they concentrated on the services it provided. Cloud manufacturing was given the green light when virtualization and resource allocation were implemented. Wu proposed incorporating 3D printing into a cloud-based system for product development and design. In addition, a cloud-based assessment methodology and a conceptual framework for a 3D printing service-oriented platform were created.

### 5.3.8 Capabilities of Cloud Computing

Cloud computing is performed on demand by using autonomous software and hardware components. One way it does this is by creating a virtualized resource pool out of previously separate resources and capabilities, making them available to clients on demand. Print quality, speed, and cost are just some of the elements that may be taken into account when allocating resources on the platform, thanks to a service evaluation and demand matching algorithm. For collaborative scheduling and planning, Wang *et al.* propose more flexible and responsive resource allocation strategies.

### 5.3.9 Hamming Distance Technique

The data was used as input into an ongoing process of creating an ML-based strategy for evaluating and choosing terminal printer services. The suggested model for evaluating services considers a number of factors, including but not limited to time, cost, quality, trust, capability, and environment. It is possible to use fuzzy numbers to quantify service quality, and the hamming distance technique and the optimization algorithm may be used to enhance the precision with which services are selected. A method for achieving quality-of-service and an evaluation model for trust in cloud manufacturing excellence were presented using the genetic algorithm.

### 5.3.10 Improving Knowledge Skills

Access to a vast library of tools is only the beginning of what an all-inclusive cloud platform has to offer. The majority of research continues to place an emphasis on technical progress, which puts future system security and reliability at risk. The risks of knowledge theft or loss are too severe in today's business environment to neglect, but the digital platform has the potential to significantly improve resource utilization and open the door for medium- and small-sized enterprises to finish their production.

It was proposed that users may be safeguarded from prying eyes by using private cloud platforms, that give users the same benefits and services as their public equivalents but are managed separately. Although there is promise in this method for mitigating risk, it requires further research and modeling to establish what those criteria and standards should be.

Cloud platforms include manufacturing as well as service data, but lack real-time management over the screen printing. It will be hard to stop or cancel a printing job after it has started, especially if the printer is connected to a cloud service. Discovering the most effective and efficient method of dealing with the massive quantities of lost time and resources is crucial.

There has been a lot of recent discussion on the importance of security in the manufacturing industry. Cyber-physical assaults have the potential to take advantage of newly discovered holes in cyber-manufacturing systems like 3D printing.

### 5.3.11 Object Detection Algorithm

Detecting harmful attacks in 3DP in real time utilizing an ML strategy based on physical data, such as the k-nearest neighbor (KNN) method, the random forest algorithm, and the object detection algorithm. Each layer-based image, defects and all, is converted into a grayscale plot. The mean, standard deviation, and the percentage of pixels that are larger than the threshold are all extracted from the grayscale values calculated.

When the cumulative distribution function of certain characteristics is unknown, KNN is utilized to determine the classification of defect regions. In real time, ML algorithms will extract features to find anomalies in faulty regions and alert the administrator. The term "anomaly detection" refers to the process of searching for data that deviates from the norm, such as when the mean value changes, the sample variance increases, or the number of pixels exceeds a certain limit.

This unsupervised learning method employs a recurrent neural network and cumulative sum algorithm to detect anomalies in a cyber-physical system with a very small number of false positives. In addition to defect classification through probability distribution function of each image layer, the random forest may produce process-based structures and use proximity to detect outliers based on the images. The vulnerability scanning approach was shown to be the most effective in both simulated and real-world experiments for identifying potential dangers.

### 5.3.12 Improvement in Manufacturing Defects

The results of the experiment prove that ML may be used to ensure 3DP security. Unfortunately, the study's findings only hold true for surfaces of components that have regular forms or patterns. It is impossible to create a detection system for a complicated building that would never incorrectly flag a good part as defective. As a result, the present state of defect detection research is only applicable to mass manufacturing of standardized components; additional improvement is needed to adapt to mass customization, which is possible only if the faults can be reliably classified.

Identification of collisions during AM operations and management of AM machines to ensure that they are free of imprint head contaminants and printing defects throughout the building process. If a worldwide system for visually regulating technical operations was developed, then there would also be visual feedback and administration. The 3D printing system undergoes a visual examination using either machine-based learning or (ii) single-point tracking. These two techniques helped increase the precision of collision detection in the 3D imprint system PROTECTS. The research makes use of methods customary of digital monitoring and reconnaissance.

## 5.4 Conclusion

In conclusion, implementing computational intelligence-based image classification for 3D printing presents several significant challenges that must be addressed to realize its full potential. The complexity of classifying digital images in the context of 3D printing, considering diverse object shapes, sizes, and materials, requires robust feature extraction methods that can capture relevant information accurately. One of the key challenges is the availability and quality of training data specific to 3D printing. With limited labeled datasets, researchers must explore transfer learning techniques and generate synthetic training data to overcome this issue. Moreover, addressing class imbalance is crucial to ensure fair representation of all object classes. The computational requirements and efficiency concerns associated with CI-based image classification in 3D printing pose additional challenges. Developing computationally efficient algorithms and optimizing the training and inference processes are necessary to achieve real-time performance, especially in resource-constrained environments. Furthermore, interpretability and explainability are critical factors for CI-based image classification in 3D printing. Understanding the decision-making process of the algorithms is essential for building user trust and enabling effective human-machine collaboration in critical applications. Despite these challenges, overcoming them through dedicated research and development efforts will lead to significant advancements in 3D printing technology. Successful implementation of CI-based image classification can automate and enhance the accuracy of 3D printing processes, enabling widespread adoption and unlocking new possibilities in various industries.

## References

1. Zhang, H. and Xia, C., Three-dimensional printing and computational intelligence: Challenges and opportunities. *IEEE Comput. Intell. Mag.*, 14, 3, 8–23, 2019.
2. Dhatterwal, J.S., Kaswan, K.S., Baliyan, A., Jain, V., Integration of cloud and IoT for smart e-healthcare, in: *Connected e-Health: Integrated IoT and Cloud Computing*, pp. 1–31, Springer International Publishing, Cham, 2022.

3. Zhou, X., Zhang, H., Chen, S., Zhang, J., Deep learning-based 3D printing quality control: Challenges and solutions. *IEEE Trans. Industr. Inform.*, 16, 3, 1651–1661, 2020.
4. Kaswan, K.S., Dhatterwal, J.S., Balyan, A., Intelligent agents based integration of machine learning and case base reasoning system, in: *2022 2nd International Conference on Advance Computing and Innovative Technologies in Engineering (ICACITE)*, IEEE, pp. 1477–1481, April 2022.
5. Li, R., Zhang, H., Li, J., Liu, Y., Transfer learning for 3D printing defect classification with limited data. *IEEE Access*, 8, 35189–35199, 2020.
6. Vasan, A. and Ramasamy, E.V., A comprehensive review on the role of machine learning in 3D printing: Challenges, applications, and future directions. *J. Manuf. Syst.*, 59, 73–90, 2021.
7. Kaswan, K.S., Baliyan, A., Dhatterwal, J.S., Jain, V., Chatterjee, J.M., Intelligent classification of ECG signals using machine learning techniques, in: *Healthcare Monitoring and Data Analysis Using IoT: Technologies and Applications, Decision and Control*, p. 412, 2022.
8. Zhu, S., Ding, H., Chen, C., Wei, Y., Zhang, J., Deep learning-based defect detection and Classification for 3D printing: A review. *Virtual Phys. Prototyp.*, 16, 3, 272–289, 2021.
9. Liu, Y., Tian, S., Wang, D., Zhang, Y., A review of machine learning algorithms for image classification in 3D printing, in: *2018 15th International Computer Conference on Wavelet Active Media Technology and Information Processing*, IEEE, pp. 427–431, 2018.
10. Kaswan, K.S., Dhatterwal, J.S., Kumar, S., Pandey, A., Industry 4.0 multiagent system-based knowledge representation through blockchain, in: *Artificial Intelligence and Industry 4.0*, pp. 93–115, Academic Press, Cambridge, Massachusetts, 2022.
11. Zhu, S., Liu, L., Wei, Y., Zhang, J., Machine learning-based porosity classification for 3D printed objects, in: *Proceedings of the 2019 IEEE International Conference on Mechatronics and Automation*, IEEE, pp. 1067–1072, 2019.
12. Huang, M., Zhang, H., Wang, Q., Wang, Y., Deep learning-based surface defect detection for 3D printed products. *Robot. Comput.-Integr. Manuf.*, 62, 101891, 2020.
13. Liu, Y., Jiang, Y., Zheng, X., Zhao, Y., . An overview of transfer learning in deep learning. *Neural Comput. Appl.*, 31, 10, 6673–6684, 2019.
14. Kaswan, K.S., Gaur, L., Dhatterwal, J.S., Kumar, R., AI-based natural language processing for the generation of meaningful information electronic health record (EHR) data, in: *Advanced AI Techniques and Applications in Bioinformatics*, pp. 41–86, CRC Press, Boca Raton, Florida, 2021.
15. Tang, X., Wang, X., Wu, X., Yin, J., Yang, G., Sun, X., Zhao, X., A review on 3D printing technology: Raw materials, fabrication, and potential applications of functional composites. *J. Mater. Sci. Technol.*, 49, 144–159, 2020.
16. Liu, Y., Wang, X., Liu, S., Hu, J., Jiang, X., Deep learning-based 3D shape classification for digital manufacturing. *J. Intell. Manuf.*, 29, 2, 401–413, 2018.

17. Dhatterwal, J.S. and Kaswan, K.S., Intelligent agent based case base reasoning systems build knowledge representation in COVID-19 analysis of recovery of infectious patients, in: *Applications of Artificial Intelligence in COVID-19*, pp. 185–209, 2021.
18. Li, H., Wu, Y., Xiao, S., Wang, H., Xie, Y., Deep learning for 3D printing: A review. *Rapid Prototyp. J.*, 26, 9, 1897–1917, 2020.
19. Mishra, A., Cao, W., Kumar, A., Ramakrishna, S., Sridhar, I., Machine learning for process-property-structure linkages in additive manufacturing: A critical review. *Mater. Today Adv.*, 8, 100089, 2020.
20. González, D., Remirez, A., Pavón, I., de la Paz, F., A review of deep learning in 3D printing: Applications, challenges, and opportunities. *IEEE Access*, 8, 110168–110188, 2020.
21. Yao, W., Gao, X., Xu, X., Shi, Y., Wang, G., Yao, J., Fang, F., A review of artificial intelligence in additive manufacturing. *J. Manuf. Syst.*, 57, 169–184, 2020.

# 6

# Role of Cybersecurity to Safeguard 3D Bioprinting in Healthcare: Challenges and Opportunities

**Venkatalakshmi S.**

*Department of Computer Science, Loyola College, Chennai, India*

***Abstract***

The act of providing medical care in an organized manner to individuals or a group is what constitutes healthcare. Health plays a pivotal role in human beings. The process of bioprinting enables the creation of living tissues, blood vessels, bones and even complete organs, which can be utilized in various medical procedures. Bioprinting is a boon to the human community and this technology has rapidly expanded and help in the production of various products ranging from car and aviation components to sophisticated medical devices and human organs. However, 3D bioprinting has advantages as well its own challenges. Research shows that 3D printed products can be counterfeited and it may lead to devastating flaws. 3D products are susceptible to attacks and threats in all phases of its use starting from its design till its implementation. Researchers find a security breach in these 3D products. Researchers opine that right from the manufacturer till the design engineers, everyone has to take a crucial role in addressing these threats. This chapter aims at making an attempt to explore and perform a systematic study on the significant role of cyber security in bioprinting. This chapter would also closely look at the solutions offered and proposed by the researchers in tackling the problem of counterfeiting the 3D products. Even though these solutions are in the infant stage, a dedicated effort is made to analyze and closely monitor the possible dimensions of solutions. One of the predominant focus of solutions will be on the materials used for bioprinting. Another important focus is on the new detection strategies to be followed for identifying the risks and threats involved in bioprinting.

*Email*: venkatalakshmi@loyolacollege.edu

E. Gangadevi, M. Lawanya Shri, Rajesh Kumar Dhanaraj and Balamurugan Balusamy (eds.)
Computational Intelligence in Bioprinting, (93–126) © 2024 Scrivener Publishing LLC

***Keywords:*** Bioprinting, 3D object, 3D bioprinting strategies, cyberthreats, additive manufacturing, blockchain, datasets in bioprinting, challenges and opportunities

## 6.1 Introduction

Can we imagine that living tissues, bone, blood vessels and human organs be printed? Yes. It's not a futuristic dream but a reality. Let us understand the forerunner 3D printing first. 3D printing involves three dimensions x(horizontal), y(vertical) and z(depth). Unlike conventional printers that print in a linear manner, 3D printers are capable of multidirectional movements, including up and down, left and right, and front and back. Additionally, instead of using ink to print on paper, they use a wide range of materials such as polymers (including plastics), metals, ceramics, and more, depending on the application. Also, an item gets printed layer by layer and hence the technology name is coined as "additive manufacturing".

Through bioprinting, it is possible to fabricate living tissues, bones, blood vessels and even complete human organs, which can be utilized for various medical procedures as shown in Figure 6.1. But, indeed it is not a simple task. Lot of complexity is involved in bioprinting since the living body has cellular complexity. Hence, Bioprinting is definitely a boon to healthcare. It has the potential to make things possible in healthcare and personalized medicine. Bioprinting could be understood as a useful extension of 3D printing.

The scope of 3D bioprinting is vast and has the potential to revolutionize several fields including medicine, tissue engineering, drug discovery

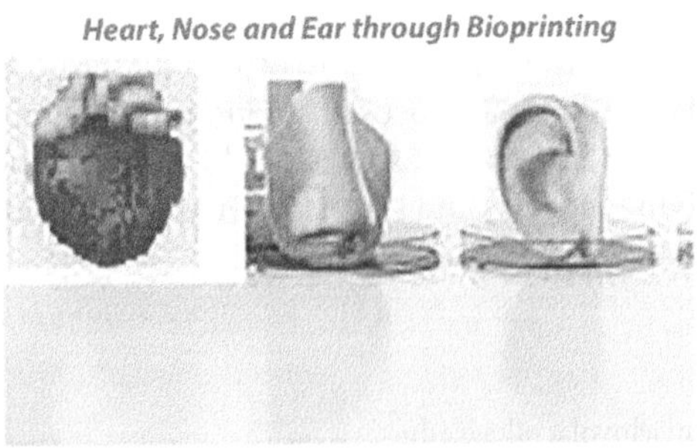


**Figure 6.1** Bioprinting.

and regenerative medicine. It provides wider opportunities for research, discovery, and clinical application.

The following are the objectives of this paper:

(a) Elaborates the creation of a 3D object, Bioprinting process and its benefits.
(b) Describes the different 3D bioprinting strategies with its advantages and limitations.
(c) Exposes to the threats and possible attacks to additive manufacturing.
(d) Gives a comprehensive survey of cyberthreats in additive manufacturing technology.
(e) Gives viable solutions and defensive mechanisms against cyberthreats.

This chapter is organized as follows:

Section 6.2 focuses on the related work in Bioprinting, its strategies, cyberthreats to additive manufacturing etc., Section 6.3 deals with the creation of a 3D object, Benefits of 3D printing, Bioprinting and the process of bioprinting. Section 6.4 discusses on the 3D bioprinting strategies, comparison between the approaches, materials used in bioprinting and its applications in diverse domains. Section 6.5 focuses on the cyberthreats posed to bioprinting, the proposed solutions in detail, combating the cybersecurity risks of bioprinting and finally a comparative survey of cyberthreats in additive manufacturing technology is stated. Section 6.6 presents the conclusion.

## 6.2 Related Work

Numerous research papers have explored the different facets of 3D bioprinting, including the various domains and applications in which this technology is utilized. The bioprinting strategies are discussed extensively. Some of the papers have also addressed about the cyberthreats/attacks to additive manufacturing.

Gu *et al.* presents the basic principles of 3D bioprinting and a detailed comparison among 3D bioprinting approaches [1].

In their work, Javaid *et al.* [2] have outlined a step-by-step process for 3D bioprinting of skin, including the imaging of burn tissue and skin. Following the printing process, the skin is subjected to post-processing, which entails tissue maturation in a bioreactor, and subsequently undergoes testing for suitability. Once tested, the skin can be readily implanted into the human body.

In her work, Abd Razak explores the legal considerations related to cybersecurity in the development of 3D bioprinting technology in Malaysia [3]. Her primary objective is to identify the areas that require improvement in the country's legal framework concerning cybersecurity, with a particular focus on 3D bioprinting.

Cyber-physical systems (CPS) refer to the integration of physical hardware with software systems through the use of a network [6]. To prevent cyberattacks, it is crucial to have a comprehensive understanding of the system's vulnerabilities and weaknesses [13].

Zhang *et al.* [8] in their paper discuss the current state of cybersecurity regulations in the medical sector and provides recommendations for improving cybersecurity measures in 3D printing.

Bello-Orgaz [16] in his paper states that there is a need for a comprehensive regulatory framework to address the cybersecurity challenges of bioprinting. The paper proposes a framework that covers the entire bioprinting process, from design and includes security measures, such as access control, encryption and data integrity checks.

El-Sayed *et al.* [17] in their paper highlight the opportunities and challenges of cybersecurity in bioprinting, including the potential for cyber-attacks and the need for secure communication channels. It recommends the framework should also consider the ethical and legal implications of bioprinting.

Fahimnia *et al.* [18] in their paper provide a systematic review of cybersecurity in additive manufacturing including bioprinting. It also provides recommendations for improving cybersecurity measures.

Chen *et al.* [19] discusses the potential for blockchain technology to enhance security and traceability in bioprinting. It also outlines the key features of blockchain technology and describes how the technology can be used to securely manage patient data.

Ramalingam *et al.* [20] present the potential cybersecurity risks associated with 3D bioprinting and propose several mitigation strategies to address these risks.

Liang-Sizhuang in his research paper on securing additive manufacturing systems has proposed Intrusion Detection System with side-channel signals for security [21]. The experiment results states that IDS with dynamic synchronization gives much better accuracy. The research paper also discusses about protecting intellectual property in AM systems. Based on the experimental outcomes, it has been demonstrated that utilizing state randomization to generate noise can effectively thwart highly sophisticated attackers.

Report [23] offers valuable insights into the present state of cybersecurity in the Additive Manufacturing industry, which includes an assessment of the industry's level of awareness and preparedness in dealing with cybersecurity threats.

## 6.3 Creation of 3D Objects and Printing

To create a 3D object, the first step involves designing a digital model using specialized software. After creating the digital model, it is then transferred to the 3D printer for further processing. Next, a critical step involves loading the chosen material into the printer, which is heated to enable smooth flow through the printer nozzle. As the printer reads the digital model, it moves its head in multiple directions—up and down, back and forth, and sideways to deposit successive layers of the chosen material, ultimately building the final product.

### 6.3.1 Benefits of 3D Printing

The most significant advantage of 3D printing is mass customization. 3D printers create more impact because of its more personalized and on demand health care products and services. Another great benefit of 3D printing is that the manufacturing and distribution processes are decentralized. These characteristics best suits for manufacturing the more personalized devices on demand.

Soft and hydrated structure
Shape customizability and clinical availability
Direct cell encapsulation
Co-culture feasibility
Combination with other delivery techniques
High *in vivo* mimetic degree (tissue analogs)

This also poses a great threat and vulnerabilities than traditionally manufactured medical devices.

### 6.3.2 Bioprinting

According to Murphy and Atala (2014), 3D bioprinting is an advanced computer-assisted technology that enables the rapid fabrication of biofunctional materials and their associated components in a layer-by-layer fashion, either on a substrate or a tissue culture dish. This cutting-edge technology allows for the creation of complex living tissues and organs that exhibit the desired 3D cellular architecture and functions.

Bioprinting can produce living tissues, bone, blood vessels and potentially whole organs for use in medical procedures, training and testing. The cellular complexity of the living body has resulted in 3D bioprinting developing more slowly than mainstream 3D printing. Bioprinting is an extension of 3D printing.

Bioprinting has proven to be an invaluable asset to the healthcare sector, with its wide range of applications in research facilities and hospitals around the world. Its potential for advancing medical research and improving patient outcomes has made it a highly sought-after technology in the field of health care as depicted in Figure 6.2.

While bioprinters operate similarly to 3D printers, their key distinction lies in the materials they use. Instead of using traditional materials such as plastic, ceramic, or metal, bioprinters deposit layers of biomaterials, which may include living cells. This unique feature allows for the creation of complex structures such as blood vessels or skin tissues.

The procedure involved in bioprinting is quite interesting. The body's various tissues are composed of distinct cell types. Therefore, to create bioink, the necessary cells are obtained from the patient and cultivated until there are sufficient cells available. Once the bioink has been produced, it is loaded into the printer for further use.

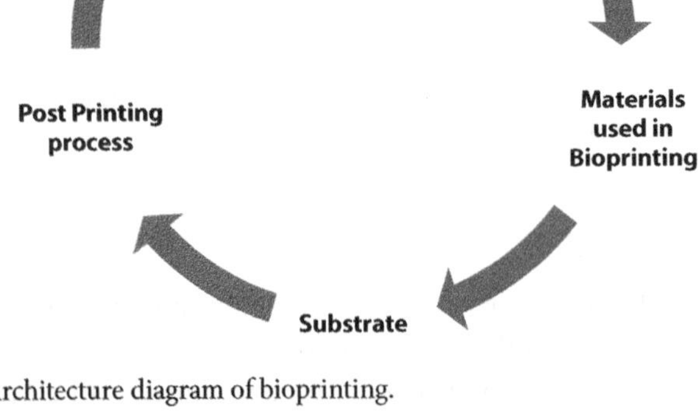


**Figure 6.2** Architecture diagram of bioprinting.

The precision printer heads are responsible for depositing the cells in their designated positions in accordance with the design. This process continues over the course of several hours, wherein the organic object is gradually constructed by stacking numerous ultra-thin layers. Since a computer design is used, it poses a security threat.

### 6.3.3 A Flow Diagram Depicting the Bioprinting Process

The process of bioprinting can be broken down into four main steps as shown in Figure 6.3:

- Data acquisition: First, data acquisition is carried out using methods such as X-ray, computed tomography, MRI or CAD software to obtain 3D models. These models are then divided into 2D horizontal slices and customized in size and orientation before being processed into particles [9].
- Material selection: The types of materials include cells, hydrogels etc., are to be chosen depending on the requirement of printed structures. The selection of bioinks is pivotal for proper printing.
- Bioprinting: The bioprinting process requires careful confirmation of printing parameters to ensure the appropriate configuration. It is crucial to closely observe the printing process to ensure accurate production of the desired product.
- Functionalization: Achieving the functions of natural tissue or organs post-printing is crucial.

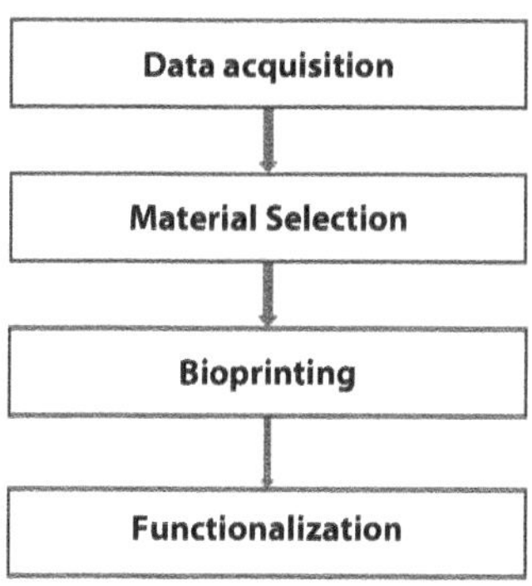


**Figure 6.3** Flow diagram depicting the bioprinting process.

### 6.3.4 Datasets Used in Bioprinting

There are several types of datasets used in bioprinting. Few examples are as follows:

(a) Medical imaging data: Creating 3D models of patient-specific tissues and organs is possible by utilizing CT scans or MRI images. These data can be processed using specialized software to create a digital model used to guide the bioprinting process.

(b) Tissue Engineering data: The integration of cells, biomaterials and growth factors is involved in the creation of functional tissues, which is known as tissue engineering. By utilizing datasets that contain information about the cell behaviour, tissue development, and extracellular matrix production, the bioprinting process can be guided to produce functional and viable tissues.

(c) Clinical data: Bioprinting is extensively used for regenerative medicine applications, such as tissue and organ replacements. Clinical data such as patient outcomes and long-term follow-up data can be collected and recorded in datasets. This could be used to evaluate the efficacy and safety of bioprinted tissues and thereby improve the bioprinting process.

In short, the use of datasets in bioprinting is crucial for optimizing the bioprinting process, ensuring the quality and safety of printed tissues thereby advancing in the field of regenerative medicine.

## 6.4 Schematic Diagram of 3D Bioprinting

The bioprinting process involves the following steps as depicted in Figure 6.4:

- Designing the 3D model
- Preparing the bioink
- Loading the bioink into the bioprinter
- Bioprinting the tissue or organ
- Post printing processing
- Quality control

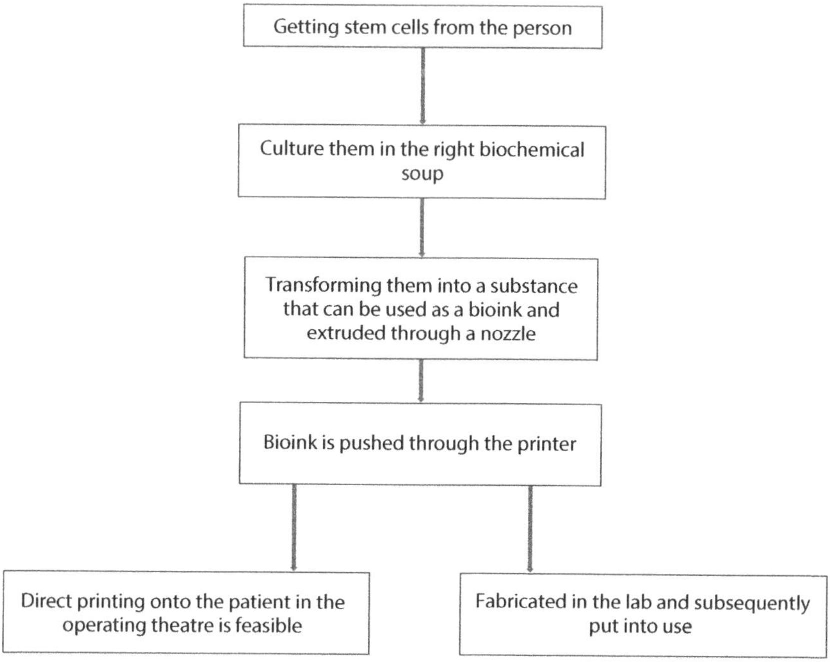


**Figure 6.4** Schematic diagram of 3D bioprinting.

The 3D bioprinting process involves the use of advanced technology and materials to create functional tissues and organs for various biomedical applications.

### 6.4.1 3D Bioprinting Strategies

There are several strategies used in 3D bioprinting strategies, each with its own strengths and limitations as shown in Table 6.1. Some of the most common strategies are listed as follows:

- Extrusion based 3D bioprinting
- Inkjet based 3D bioprinting
- Laser assisted 3D bioprinting
- Stereolithographic based 3D bioprinting
- Magnetic bioprinting

Extrusion-based is the most commonly used bioprinting technique. In this method a bio-ink made up of cells and other biological materials is

**Table 6.1** Comparison among 3D bioprinting approaches.

| Type of printing | Description | Advantages | Disadvantages | Domain |
|---|---|---|---|---|
| Extrusion based bioprinting | Expulsion of bioink through a nozzle | (i)Bioinks with high cell density can be utilized, enabling printing at a higher density.<br>(ii) A vast array of biocompatible materials can be printed due to the technology's versatility.<br>(iii) Converting commercial plotters or 3D printers into extrusion-based bioprinters is a straightforward process. | (i)The precision of extrusion-based printing is constrained to 100 μm, which is less than that of biological techniques.<br>(ii)Cell structure may be distorted because of undue stress during printing. | Tissue and Organ engineering research. |
| Inkjet based bioprinting | Expulsion of successive drops of bioink on a substrate under thermal or electromagnetic effect | (i)Easily available<br>(ii) cheap and high speed technique | (i)Lack of precision<br>(ii)Low viscosity bioink is required.<br>(iii)Cells are susceptible to mechanical or thermal harm during the printing procedure. | Tissue engineering and therapeutics research. |

*(Continued)*

**Table 6.1** Comparison among 3D bioprinting approaches. (*Continued*)

| Type of printing | Description | Advantages | Disadvantages | Domain |
|---|---|---|---|---|
| Laser assisted bioprinting | The bioink and cell suspension situated at the bottom of the ribbon are vaporized by a laser pulse, which then propels them onto the receiving substrate. | (i)Exceptional accuracy and resolution can be achieved.<br>(ii)The absence of a nozzle in the printing process eliminates the issue of biomaterials obstructing the nozzle. | (i)The process is costly and time consuming.<br>(ii)The availability of hydrogel materials appropriate for laser assisted bioprinting is limited. | Used to fabricate different tissue constructs for tissue engineering applications. |
| Stereolithography based bioprinting | Photocurable bioink is cured in a layered manner by means of light. | (i) The process provides a high level of accuracy in fabrication while minimizing printing time. | (i)Lengthy post-processing, highly complicated process. | It has applications in generating artificial tissues and organs, as well as depositing living cells at a microscale level. |
| Magnetic based bioprinting | Cells are suspended in a magnetic bio-ink and deposited onto a substrate using a magnetic field. | (i)Precise control over the placement of the cells. | (i)Limited by the type of cells that can be used. | Tissue Engineering, Drug discovery, Cancer research and Neurobiology. |

extruded through a nozzle onto a substrate layer-by-layer for the creation of a 3D structure. The bio-ink may be curated after each layer is deposited to maintain the structural integrity of the final product.

In Inkjet-based bioprinting, droplets of a bio-ink are deposited onto a substrate using a printhead. This can achieve high resolution and precision.

In Laser-based bioprinting, laser is used to create tiny droplets of a bio-ink and then deposited onto a substrate to create a 3D structure. Despite the limitation of the materials used in the bio-ink, this technique allows for precise control over the shape and size.

Stereolithography-based bioprinting used a photosensitive resin that is cured using a light source to create a 3D structure. The cured structure can then be further processed to create a functional tissue.

### 6.4.2 Comparison Among the 3D Bioprinting Approaches

Different bioprinting methods are discovered appropriate to the different applications and each of these methods possess corresponding advantages and disadvantages.

### 6.4.3 Materials Used in Bioprinting

Biomaterials are substances, either natural or synthetic as shown in Figure 6.5, utilized for the repair or replacement of organs. They fall under four categories: metals, polymers, ceramics or composites, with thermoplastic polymers being the most common in 3D bioprinting. Polymeric bioinks offer numerous advantages, including biocompatibility, degradability and low cost. Hydrogels are particularly well suited for tissue engineering due to their ability to retain large amounts of water within their structure.

### 6.4.4 Bioprinting in Diverse Domains

As shown in Figure 6.6, bioprinting is used in several domains as follows:

(a) Organ and tissue printing
The rapid production of organs is a crucial application of 3D bioprinting. This technology is also utilized in dentistry for the printing of organs and tissues. The most important use case of bioprinting is making replacement organs [7].

(b) Muscle tissue printing
Skeletal muscle tissue engineering (SMTE) is focused on repairing damaged skeletal muscles and bioprinting is being explored as a promising tool for achieving this goal.

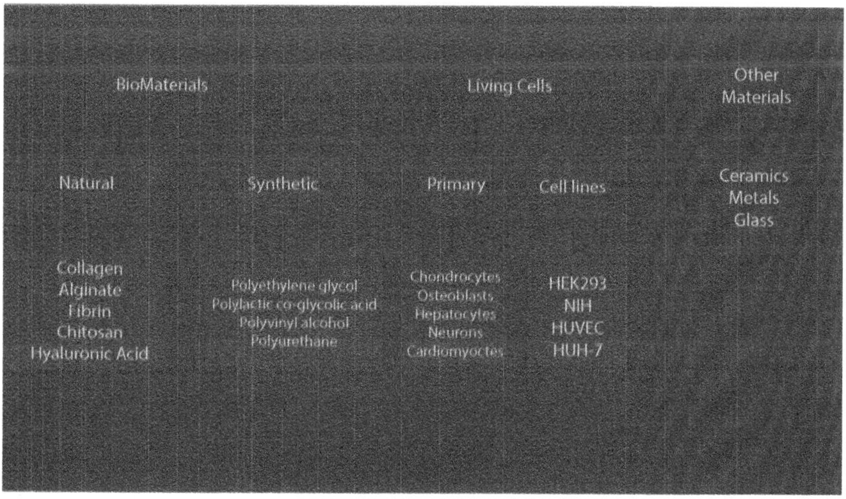


**Figure 6.5** Classification of materials in bioprinting.

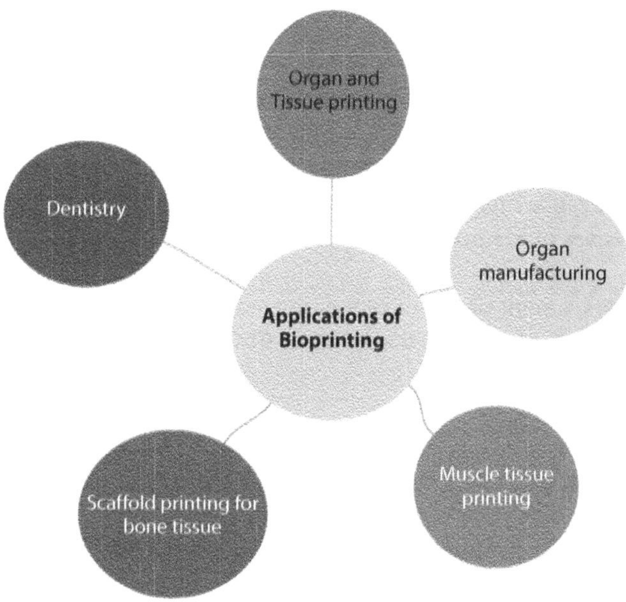


**Figure 6.6** Bioprinting in diverse domains.

(c) Scaffold printing for bone tissue
Three-dimensional bioprinted scaffolds are the foundation of bone tissue engineering (BTE), which has emerged as a promising method for bone regeneration and treatment.

(d) Organ manufacturing
Bioprinting organs can well tackle the severe shortage of organ donor supply. Organ manufacturing is a boon to the medical field to replace defective/failure organs. It is a complex interdisciplinary field that requires expertise and huge number of talents in multiple disciplines namely, biology, chemistry, physics, mechanics, materials, computer science etc., Production of bioartificial organs is possible through 3D bioprinting by automatic layer-by-layer deposition methods [12].

A crucial aspect of 3D bioprinting technology is the utilization of living cells, polymeric hydrogels, and bioactive agents, known as "bioinks", to construct bioartificial organs based on a computer-aided design model. Furthermore, multiple cell types can be encapsulated in distinct polymeric hydrogels and deposited simultaneously.

(e) Dentistry
3D printers are capable of producing precise orthodontic models, restorations, dentures, aligners, retainers, crowns, bridges and other dental prosthetics. The material used in the printing process varies according to the treatment to be done. 3D bioprinting is extremely useful in dental treatments because high customization is required.

The use cases of 3D printing in dentistry are

- Replacement or repair of a damaged tooth
- Creation of mouth guards, teeth straightening aligners
- Surgical tools

## 6.5 Cyberthreats Posed to Bioprinting

Biopriting is very much advantageous in many fields especially in medical field. Given that medical devices are no longer standalone components, cybersecurity is an essential aspect that must be incorporated into them [14]. Hence, it becomes a necessity to entirely revamp the risk management processes so that any kind of vulnerabilities could be faced. Therefore, as we envisage serious threats starting from the design of the 3D products till the manufacturing stage, it becomes vital that several precautions must be in place to face any tough and undesirable situations as pointed in Table 6.2.

It becomes an essential part to identify the security vulnerabilities and propose viable solutions to mitigate the cybersecurity risks anticipated. This is not easy as stated since securing medical devices against cyber threats is a complex and multifaceted challenge that grows every day.

According to the RegData industry Regulation index, the medical industry is among the top ten most regulated industries due to its vast amounts of data and ever-changing requirements. Additionally, the medical manufacturing market is identified as one of the top five industries that are highly targeted and vulnerable to cyber attacks.

As the medical industry digitizes its supply chain and manufacturing processes, it is crucial to carefully address some key challenges as stated in Table 6.3.

To cut costs, some organizations outsource the printing of software-designed products to external facilities. But the serious issue with this is that the firmware in the printers could be hacked. If the firmware is hacked, it results into a defective product being printed. The important concern is

**Table 6.2** Type of risks and its consequences.

| Type of risk | Consequences |
|---|---|
| IP risk | If hackers gain access to the data, it can lead to potential loss of revenue for the Intellectual Property (IP) owner. |
| Production risk | The quality and performance of parts can be compromised when they are produced by any owner of a 3D printer. |
| Traceability risk | Authenticity of a physical part can be questioned when there is no clear connection between it and its digital twin. |
| Liability risk | Loss of control over data or the production process can lead to legal liability for manufacturers, particularly in the case of defects in their product design. |
| Confidentiality risk | Ensuring the confidentiality and privacy of personal health records is of utmost importance. Confidentiality and data integrity could be a great matter of concern during the production of individualized/customized medical devices. |

that the products will not look defective on the outside but may have fractures or holes inside them.

The 3D printer is equipped with an extruder or arm that extrudes material to create an object. It has been observed that just by observing the motion of the extruder, using sensors, it could be checked whether the print process follows the design meant or not.

Another novel method is to examine the object to check whether it was printed correctly. In a process called "nanoparticle-enabled 3D printing", tiny gold nanoparticles are incorporated into the 3D printing process. These nanoparticles are mixed into the printing material and become part of the printed object. Once the object is printed, scanning reveals whether the object is printed properly or has any holes or defects.

Design stage:
It is highly possible to steal the design since the complete product is digitally manufactured. Also, researchers say that some of the designs can even be copied just following the vibration of a 3D printer at work. Tampering the significant component of a design is also equally possible.

Manufacturing stage:
It is also found out that not all defects in the 3D product could be identified using present quality control methods and techniques. Design data corruption is also a matter of concern that can pose serious health risks. Even the change in the orientation of extrusion head may result in a different structure with different properties.

Indeed, while the adoption of advanced medical device technology and 3D bioprinting can bring many benefits to healthcare facilities, it is important to recognize that these devices also have inherent vulnerabilities that can create pathways for cyber threats and attacks. The protection of sensitive patient data and the prevention of unauthorized access to medical devices are critical concerns that need to be addressed through effective cybersecurity measures. These challenges require a multifaceted approach [15] that considers the entire lifecycle of medical devices, from design and development to use and disposal.

There are certain common vulnerabilities that can lead to threats:

Hardcoded administrative passwords:
Hardcoded passwords are to be changed otherwise the devices could be prone to attack since a cyber actor can guess and identify the password and may gain administrator access to the device.

Unencrypted data transmitted in the network:
Interception of data transmitted through a wireless network is possible if it is not properly encrypted.

Traditional security measures that are not updated:
Antivirus software, Firewall and Intrusion detection systems are very much essential to safeguard the medical devices. But, if it is not up-to-date, this could incapacitate the system.

### 6.5.1 Challenges and Opportunities of Cybersecurity in Bioprinting

**Table 6.3** Challenges and opportunities.

| Challenges | Opportunities |
|---|---|
| Data security | Enhanced security measures |
| Intellectual property protection | Collaborative research |
| Cyberattacks | Cybersecurity workforce development |
| Regulatory Compliance | Advancement in cybersecurity technology |
| | Opportunities for innovation |

### 6.5.2 Proposed Solutions

In general, the hospitals and healthcare facilities shall always be vigilant in protecting their network and mitigating the risk involved in maintaining the patient health care records.

In order to protect the additive manufacturing systems from cyber-attacks, usage of Intrusion detection system is highly recommended. Researchers have proposed a range of Intrusion Detection Systems (IDSs) that use side-channel signals [4] in recent years. These signals are physical signals that are correlated with the state of the additive manufacturing system, such as the acoustic sound or the electromagnetic waves emitted by a 3D printer during printing. As a result, it is possible to perform intrusion detection by analysing the correlation between a side-channel signal and the state of a 3D printer as stated in Table 6.4.

According to FDA safety communication, it recommends that the medical device manufacturers should take necessary measures to ensure that their devices are equipped with appropriate security measures to minimise the risk of failure due to cyber-attacks. To achieve this, manufacturers must consider factors such as

(i) The intended use of the device.
(ii) The presence and purpose of electronic data interfaces.
(iii) The intended environment.
(iv) The type of cybersecurity vulnerabilities that are present.
(v) The likelihood that the vulnerabilities will be exploited
(vi) The probable risk of patient harm due to a cybersecurity breach.

It is always recommended to address the cybersecurity challenges during the design stage. The device's safety and efficacy are to be taken care in this stage itself.

Functionality of the device:
While certain security controls such as multi-layered authentication are needed to resist unauthorized access, there are possibilities that these measures can also probably slow down the device's functionality and reduce its shell life.

Periodic updates and maintenance
To ensure their security, connected medical devices typically require regular updates and patches.

It is highly significant that security has to be included in the medical device's architectural risk assessment. Device manufacturers must ensure that their products do not introduce additional vulnerabilities to the network of healthcare facilities. This requires careful consideration of the device's connectivity, access controls and security protocols. Regular security assessments and updates may be necessary to address new threats and vulnerabilities.

### 6.5.3 Combating the Cybersecurity Risks of 3D Bioprinting

As 3D printing and bioprinting surges, cyberattacks and threats against this technology are a real threat. Since these technologies are cloud based, they are susceptible to malicious attacks by hackers. 3D printers are at the risk of hacking attacks that leads to design leaks, copyright infringements and loss of sales and reputation for a company.

These threats are due to the fact that 3D printers are connected to the internet and the shared use of manufacturing-as-a-service networks. The data are susceptible to malicious attacks whose intention is to obtain the digital design files, steal intellectual property (IP), create a counterfeit design, manipulate output traceability, alter the integrity of designs and introduce design modifications.

Researchers suggest the following steps to combat cybersecurity risks in 3D printing:

Risk assessment: It is the identification of cybersecurity risks that could affect the productivity of 3D printing. A proper risk assessment helps us to identify the risks and understand their impact.
Network security: Network has to be secured so that any malicious attacks could be monitored, detected and prevented.

Guidelines to ensure the network is secure.

- Perform a network audit.
- Limit access to your printer.
- Use strong passwords.
- Update antivirus software.
- Segment and segregate your network.
- Use VPNs.

Encryption of files: Files can be encrypted with strong keys for more security.

Regular inspection: Regular security inspection is needed to check the systems periodically. Design inspection of 3D printers, code inspection, deployment inspection of printing materials helps to combat cybersecurity risks in bioprinting in a great way.

Expert consultation: Consulting the experts help to implement the best security practices to prevent cyberthreats and to optimize the 3D printing process.

### 6.5.4 Blockchain Technology and Bioprinting

Blockchain technology has an inherent connection to cybersecurity. Storage of information, transaction management, trust establishment are done in a different way using blockchain technology as shown in Figure 6.7. Hence, it becomes very conducive to the environments which demands high security.

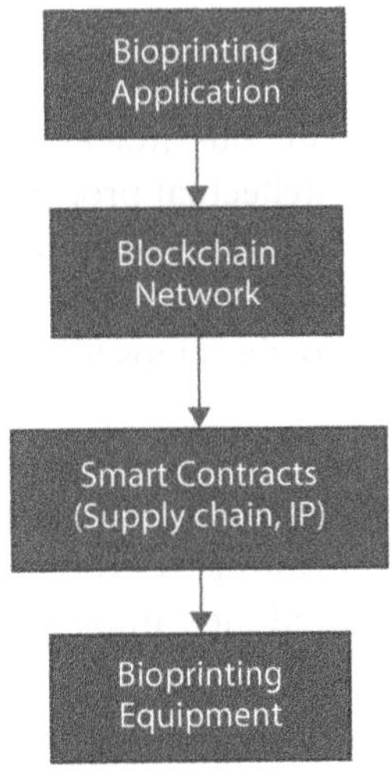


**Figure 6.7** Blockchain and bioprinting.

The utilization of Blockchain technology in the creation of a Keyless Signature Infrastructure (KSI) presents an alternative to the conventional public key infrastructure (PKI). The KSI employs asymmetric encryption and public keys to secure data.

Traditional systems' reliance on simple logins and a centralized architecture presents significant vulnerabilities. Blockchain takes the responsibility of strong authentication thereby mitigating the risks in security breach. This system utilizes a distributed public key infrastructure to verify the identity of both users and devices. This technology gives each device a particular SSL certificate thereby avoiding passwords. Due to the management of data on the Blockchain, SSL certificates are highly secure against fraudulent activities by malicious hackers.

Hence, it is high time to explore the advantages and benefits of blockchain technology in bioprinting too.

### 6.5.5 A Comparative Survey of Cyberthreats in Additive Manufacturing Technology

**Table 6.4** Comparative survey of cyberthreats in AM.

| Title | Motivation | Keypoints | Conclusion | Future directions | References |
|---|---|---|---|---|---|
| State of Security Awareness in the AM industry: 2020 Survey | The widespread adoption of additive manufacturing (AM) in various applications has drawn increased attention to its security. However, a significant gap exists between AM manufacturing and its security measures. | Surveyed the AM community, asking total 11 AM security related questions aiming to discover the existing concerns and expectations. | The author's conclusion, based on a survey of 53 participants, suggests that while the responses are somewhat representative of the current state of the AM industry, the industry is not adequately prepared to prevent or detect attacks that are specific to AM. | | [23] |

(*Continued*)

**Table 6.4** Comparative survey of cyberthreats in AM. (*Continued*)

| Title | Motivation | Keypoints | Conclusion | Future directions | References |
|---|---|---|---|---|---|
| Cyber-physical vulnerabilities in AM: A case study attack on the .STL file with human subjects | AM's digital thread is often touted as a key advantage, as it enables speedy communication, iteration, and sharing of both the design model and the physical object it represents. However, this also creates new avenues for cyberattacks. | The study investigated the possible ways in which the AM process chain could be targeted for cyber-attacks. It delved into the consequences of cyber-physical attacks and also explored the different methods of detecting such attacks. | A case study was conducted to evaluate the ability of human subjects to detect and diagnose a cyber physical attack on the .STL file of a test specimen, in order to explore the potential implications of such attacks. | The paper puts forth suggestions aimed at preventing and identifying cyber physical attacks on AM procedures. | [6] |

(*Continued*)

**Table 6.4** Comparative survey of cyberthreats in AM. (*Continued*)

| Title | Motivation | Keypoints | Conclusion | Future directions | References |
|---|---|---|---|---|---|
| Cyber security challenges in 3D Bioprinting: Legal analysis | To explore the cyberthreats to 3D bioprinting legally | The article examines the legal aspects of cybersecurity in the development of 3D bioprinting in Malaysia and provides recommendations for enhancing the legal framework to address cybersecurity concerns, specifically in the context of 3D bioprinting. | Recommendation to improve the security legal framework pertaining to 3D bioprinting. | | [3] |

(*Continued*)

**Table 6.4** Comparative survey of cyberthreats in AM. (*Continued*)

| Title | Motivation | Keypoints | Conclusion | Future directions | References |
|---|---|---|---|---|---|
| Securing cyber-physical additive manufacturing systems by *in-situ* process authentication using streamline video analysis | Cyber-attacks in additive manufacturing (AM) cyber-physical systems (CPS) have the potential to severely impact the design of the AM component, ultimately leading to compromised mechanical properties and functionality. | Ensuring that the AM part is fabricated as intended can be achieved through *in-situ* process authentication. Cyber-physical attacks on AM processes can often be observed as changes in the printing path, which can be detected using an optical imaging system during the printing process. | The proposed methodology involves using image texture analysis of *in-situ* videos to authenticate the AM process. This is done by characterizing the layer-wise distribution of segmented textures' geometric features as the layer-wise texture descriptor tensor (LTDT). To reduce the high dimensionality and sparsity of the LTDTs, the multilinear principal component analysis (MPCA) algorithm is applied. | The proposed method has been shown to be highly accurate, with a detection rate of over 95%. This indicates that the method is effective in detecting alterations in the printing path that may lead to changes in the AM part's properties. | [5] |

(*Continued*)

**Table 6.4** Comparative survey of cyberthreats in AM. (*Continued*)

| Title | Motivation | Keypoints | Conclusion | Future directions | References |
|---|---|---|---|---|---|
| | | Implementing *in-situ* authentication can prevent significant geometric changes and mechanical property compromises in AM parts, thereby enhancing the overall security of the AM process. | The Hotelling T2 control charting technique is then used for alteration detection based on the extracted low-dimensional layer-wise features. The efficacy of this approach is evaluated through case studies involving a fused filament fabrication (FFF) process. | | |

(*Continued*)

**Table 6.4** Comparative survey of cyberthreats in AM. (*Continued*)

| Title | Motivation | Keypoints | Conclusion | Future directions | References |
|---|---|---|---|---|---|
| Architecture for preventing and detecting cyber-attacks in cyber-manufacturing system | A Cyber-Manufacturing System (CMS) is a convergence of operational and informational components within manufacturing processes, aimed at enhancing productivity. Its integration enables prompt product delivery, remote access to manufacturing processes, and real-time process analysis. | Cyber-Manufacturing Systems (CMS) are vulnerable to cyberattacks that can disrupt the manufacturing process, cause product alterations, machine failures, or even fatal accidents. These attacks may be carried out intentionally or unintentionally, creating potential security risks for the system. | The researchers conducted a comprehensive review of existing literature to examine the various methods available for detecting and preventing cyber physical attacks in CMS. They proposed a 5-layer architecture that integrates detection and prevention mechanisms to enhance the security of CMS against such attacks. | | [10] |

(*Continued*)

**Table 6.4** Comparative survey of cyberthreats in AM. (*Continued*)

| Title | Motivation | Keypoints | Conclusion | Future directions | References |
|---|---|---|---|---|---|
| Detecting Sabotage attacks in Additive manufacturing using Actuator power signatures | As the adoption of additive manufacturing (3D printing) increases globally, the technology is being used to produce safety-critical system components. However, due to its reliance on computers, there is a growing concern that the AM process may be susceptible to tampering, potentially compromising the mechanical properties of printed parts. | | This paper proposes a new method for identifying sabotage attacks in manufacturing systems. The method involves continuous monitoring of the current supplied to each actuator during the manufacturing process, and detecting any anomalies that may indicate tampering. | | [11] |

(*Continued*)

**Table 6.4** Comparative survey of cyberthreats in AM. (*Continued*)

| Title | Motivation | Keypoints | Conclusion | Future directions | References |
|---|---|---|---|---|---|
| Cyber-physical watermarking with Inkjet edible bioprinting | | | The article explores the use of print-camera watermarking as a cybersecurity solution to prevent counterfeiting and provide authentication features for medication at the dosage level. The proposed method combines digital watermarking and edible printing to ensure that the watermark is both edible and printable, while also being imperceptible, tamper-resistant, robust, and user-friendly. | The on-dose printing method proposed in this paper has various potential applications, such as the development and production of single-unit packages or unit-dose packages in hospital pharmacy settings. By lowering the risk of dispensing errors, improving inventory tracking, enhancing security, and minimizing labour costs, this method can improve patient safety and reduce healthcare costs. | [22] |

(*Continued*)

**Table 6.4** Comparative survey of cyberthreats in AM. (*Continued*)

| Title | Motivation | Keypoints | Conclusion | Future directions | References |
|---|---|---|---|---|---|
| | | | | The edible watermarking technique can also allow patients to play an active role in preventing counterfeit medicines. Additionally, this method can be used for other cryptographic and security protocols that require immediate obliteration of information representation and data storage after being accessed. | |

(*Continued*)

**Table 6.4** Comparative survey of cyberthreats in AM. (*Continued*)

| Title | Motivation | Keypoints | Conclusion | Future directions | References |
|---|---|---|---|---|---|
| Securing Additive Manufacturing Systems from Cyber and Intellectual Property Attacks | (a) An increased risk of cyberattacks on systems. | The proposed intrusion detection systems (IDSs) rely on side-channel signals, which are physical signals that are associated with the state of the AM system, such as acoustic sound or electromagnetic waves emitted during a 3D printing process. These signals can be analyzed to detect intrusions, as there is a correlation between the side-channel signals and the state of the 3D printer. | The study proposed the use of dynamic synchronization for improving intrusion detection systems (IDS) that use side-channel signals in additive manufacturing (AM) systems. The aim was to find corresponding points between two signals in real-time, demonstrating its necessity. The study conducted a series of experiments using two printers, performing a total of 302 benign printing processes and 200 malicious printing processes. | | [21] |

(*Continued*)

**Table 6.4** Comparative survey of cyberthreats in AM. (*Continued*)

| Title | Motivation | Keypoints | Conclusion | Future directions | References |
|---|---|---|---|---|---|
| | | | The results showed that existing IDSs using side-channel signals in AM systems had an accuracy range of 0.50 to 0.88, while the proposed IDS with dynamic synchronization achieved a significantly higher accuracy of 0.99. | | |
| | (b) Cyberattacks to steal intellectual property in AM systems. | The paper examines the vulnerability of additive manufacturing (AM) to optical side-channel attacks, which are easier to perform than acoustic side-channel attacks. | The researchers proposed a solution to defend against optical side-channel attacks in additive manufacturing by introducing an optical noise injection method. | | |

(*Continued*)

**Table 6.4** Comparative survey of cyberthreats in AM. (*Continued*)

| Title | Motivation | Keypoints | Conclusion | Future directions | References |
|---|---|---|---|---|---|
| | | The optical side-channel signal is the video of the printing process, and the coordinates of the printhead in each frame of the video can be recognized using a modified deep neural network called ResNet50. To defend against the optical side-channel attack, the paper proposes the optical noise injection method. | The method involves using an optical projector to inject artificial optical noise onto the printing area, making it difficult for an attacker to recover the printing path. Three noise generation algorithms were proposed, and the experiment results demonstrated that the noise generated using state randomization was effective in defeating the advanced attacker. | | |

## 6.6 Conclusion

Bioprinting has gained immense popularity in the recent years accompanied also by a rise in cybersecurity risks. In this chapter, we have analyzed about 3D printing, bioprinting, its strategies, applications, the cyberthreats posed to bioprinting and the proposed solutions. It is evident that 3D bioprinting has a great future in healthcare. Despite its usefulness there are several risks from the design stage till the manufacturing stage. Therefore, all the stakeholders in the manufacturing of a bioprinting product are to play a responsive role. Since 3D printers and bioprinting processes are mostly cloud based, they are most likely to be prone to cyberattacks. Hence, it is highly essential to implement the best cybersecurity processes in this additive manufacturing industry. A clear attempt has been made in this paper to understand all the above concepts. A thorough study is also made on the cyberthreats posed to additive manufacturing and the possible solutions. If appropriate preventive measures are properly taken by the industry, bioprinting will definitely reach greater heights in the service of mankind.

## References


1. Gu, Z., Fu, J., Lin, H., He, Y., Development of 3D bioprinting: From printing methods to biomedical applications. *Asian J. Pharm. Sci.*, 15, 5, 529–557, September 2020.
2. Javaid, M. and Haleem, A., 3D bioprinting applications for the printing of skin: A brief study. *Sens. Int.*, 2, 100123, July 2021. https://doi.org/10.1016/j.sintl.2021.100123.
3. Abd Razak, S.S., Cyber security challenges in 3D bioprinting: Legal analysis. *UiTM International conference on Law & Society*, 2022.
4. Agarwal, S., Saha, S., Balla, V.K., Pal, A., Barui, A., Bodhak, S., Current developments in 3D bioprinting for tissue and organ regeneration-a review. *Front. Mech. Eng.*, 6, 589171, 1–22, October 2020.
5. Al Mamun, A., Liu, C., Kan, C., Tian, W., Securing cyber-physical additive manufacturing systems by *in-situ* process authentication using streamline video analysis. *J. Manuf. Syst.*, 62, 429–440, 2022.
6. Sturm, L.D., Williams, C.B., Camelio, J.A., White, J., Parker, R., Cyber-physical vulnerabilities in additive manufacturing systems: A case study attack on the .STL file with human subjects. *J. Manuf. Syst.*, 44, Part 1, 154–164, 2017.
7. Mota, C., Camarero-Espinosa, S., Baker, M.B., Wieringa, P., Moroni, L., Bioprinting: From tissue and organ development to *in vitro* models. *Chem. Rev.*, 120, 19, 10547–10607, 2020.
8. Zhang, Z., Kiani, H., Alam, M., Cybersecurity Risks of 3D printing in the medical sector: A review. *Virtual Phys. Prototyp.*, 15, 1, 1–13, 2020.

9. Gu, Z., Fu, J., Lin, H., He, Y., Development of 3D bioprinting: From printing methods to biomedical applications. *Asian J. Pharm. Sci.*, 15, 5, 529–557, September 2020.
10. Prasad, R. and Moon, Y., Architecture for preventing and detecting cyber attacks in cyber-manufacturing system. *IFAC-Pap.*, 55, 10, 2246–2251, 2022.
11. Gatlin, J., Belikovetsky, S., Moore, S.B., Solewicz, Y., Elovici, Y., Yamposkiy, M., Detecting sabotage attacks in additive manufacturing using actuator power signatures. Special section on Additive manufacturing security. *IEEE Access*, 1, 1–12. 2019.
12. Javaid, M. and Haleem, A., 3D bioprinting applications for the printing of skin: A brief study. *Sens. Int.*, 2, 100123, 2021.
13. Kumara, R., Rogall, C., Thiede, S., Herrmann, C., Sangwan, K.S., Development of a decision support system for 3D printing processes based on cyber physical production systems. ScienceDirect *Proc. CIRP*, 98, 348–353, 2021.
14. Zeltmann, S.E., Gupta, N., Tsoutsos, N.G., Maniatakos, M., Rajendran, J., Karri, R., Manufacturing and security challenges in 3D printing. *JOM*, 68, 7, 1872–1881, 2021.
15. Yampolskiy, M., King, W.E., Gatlin, J., Belikovetsky, S., Brown, A., Skjellum, A., Elovici, Y., Security of additive manufacturing: Attack taxonomy and survey. *Addit. Manuf.*, LLNL-JRNL-738793, 21, 431–457, 2017.
16. Bello-Orgaz, G., Hernandez-Castro, J.C., Camacho, D., Bioprinting security: The need for a comprehensive regulatory framework. *Front. Bioeng. Biotechnol.*, 7, 311, 2019. https://doi.org/10.3389/fbioe.2019.00311.
17. El-Sayed, M. and Ghaith, A., Cybersecurity in bioprinting: Opportunities and challenges. *Front. Bioeng. Biotechnol.*, 8, 864, 2020. https://doi.org/10.3389/fbioe.2020.00864.
18. Fahimnia, B., Sarkis, J., Davarzani, H., A systematic review of cybersecurity in Additive Manufacturing: Recommendations for secure bioprinting. *Int. J. Prod. Res.*, 59, 8, 2324–2347, 2021. https://doi.org/10.1080/00207543.2020.1853718.
19. Chen, Y., Liu, Y., Zhao, X., Yang, S., Wu, T., Blockchain technology in Bioprinting: Enhancing security and traceability. *J. Biomed. Mater. Res. Part B Appl. Biomater.*, 109, 5, 617–625, 2021. https://doi.org/10.1002/jbm.b.34726.
20. Ramalingam, S.P., Sivanandan, S.S., Selvamani, K., Cybersecurity risks and mitigation strategies in 3D Bioprinting. *J. Med. Syst. Springer*, 45, 7, 91, 2021.
21. Liang, S., *Securing Additive Manufacturing Systems from Cyber and Intellectual Property Attacks*, Georgia Institute of Technology, Proquest LLC, Ann Arbor, Michigan, USA, 2021.
22. Jeon, H.-J., Leem, J.W., Ji, Y., Park, S.M., 3D printing of multiscale, multifunctional and hierarchical structures based on carbon nanotube hybrid materials. *Adv. Funct. Mater.*, 31, 41, 2112479, 2021. https://doi.org/10.1002/adfm.202112479.
23. Yampolskiy, M., Bates, P., Siefi, M., Shamsaei, N., *State of Security Awareness in the AM Industry: 2020 Survey*, NCMS, Ann Arbor, Michigan, USA, https://www.ncms.org/wp-content/uploads/2020/12/State-of-Security-Awareness-in-the-AM-Industry-2020-Survey-Report-Final.pdf.

# 7

# Legal and Bioethical View of Educational Sectors and Industrial Areas of 3D Bioprinting

**Pothys varan S.[1], Balachander S.[2] and Ashwini S.[2*]**

[1]*School of Law, Christ (Deemed to be University), Bengaluru, Karnataka, India*
[2]*Department of Computing Science & Engineering, SRM Institute of Science and Technology, Chennai, Tamil Nadu, India*

***Abstract***

Recent advancements in three-dimensional printing (3D printing) within the medical field, particularly in the realm of 3D bioprinting, have shown tremendous potential in transforming various medical therapies, offering new approaches to treat organ failure and injury. However, amidst this optimism, several significant ethical and legal challenges remain unresolved before the application and transplantation of 3D bioprinted technology and organs in human subjects can become a reality. This chapter focuses on exploring the ethical and legal constraints associated with 3D bioprinting technology from both educational and industrial perspectives, recognizing their crucial roles as cornerstones for future applications. Furthermore, the analysis of 3D bioprinting technology will be conducted through the lens of the fundamental medical ethics principle, "Primum non nocere"; First, do no harm. Moreover, the pressing need for effective and timely standalone laws to regulate the subject of 3D printing is emphasized. This urgency arises from the grave concerns posed by the future implications of this technology on India's scientific research and medical practice. The aim of this paper is to provide a comprehensive examination of the ethical and legal challenges posed by 3D bioprinting technology. By considering both educational and industrial perspectives, this research seeks to shed light on the complexities surrounding the application and transplantation of 3D bioprinted organs. Additionally, the analysis through the principle of "Primum non nocere" will contribute to the understanding of the ethical implications inherent in this innovative technology. Ultimately,

**Corresponding author*: as9792@srmist.edu.in

E. Gangadevi, M. Lawanya Shri, Rajesh Kumar Dhanaraj and Balamurugan Balusamy (eds.)
Computational Intelligence in Bioprinting, (127–156) © 2024 Scrivener Publishing LLC

this study advocates for the formulation of appropriate regulations and guidelines through the implementation of effective standalone laws, ensuring the responsible development and utilization of 3D printing technology in the realm of scientific research and medical practice in India.

***Keywords:*** 3D bioprinting, legal, medical practices, analysis

## 7.1 Introduction

In recent times, there has been a rise in the field of bioprinting, which involves the use of additive manufacturing techniques to create three-dimensional structures using a combination of living and non-living materials. The primary objective of bioprinting is to develop constructs that mimic the functionality of natural tissues and organs. This process utilizes computer-aided design software to design and manufacture bioengineered structures that find applications in various fields such as regenerative medicine, tissue engineering, pharmacokinetics, medical research, drug discovery, and cell biology studies [1]. Unlike traditional 3D printing methods, bioprinting introduces a groundbreaking element by incorporating living cells into a specialized substance known as bio-ink. This presents unique challenges as it requires maintaining the viability of the living cells while constructing three-dimensional structures that closely resemble natural tissues and organs.

Both the industrial and academic sectors have recognized the significant potential of bioprinting techniques, leading to a surge in research and technological advancements. The field of biomanufacturing has made remarkable progress, with the development of specialized 3D bioprinters [2] and innovative biomaterials known as bio-inks [3].

A bioprinter [4] is a specific type of 3D printer that constructs biological tissue structures by layering living cells onto a substrate. These bioprinters typically utilize bio-inks, which are soft biomaterials infused with living cells, to create intricate biological constructs. Another approach involves the use of dissolvable materials to provide support and protection during the printing process.

However, despite the extensive focus on techniques and classifications of bio-inks in recent bioprinting research [7], there remains a lack of comprehensive legal and bioethical analyses. This includes an examination of academic papers, patents, and industry players [5], aiming to understand the current distribution of key stakeholders in academia and industry, as well as the primary areas of research being explored [6].

Against this backdrop, the purpose of this paper is to capture the attention of scientists [1] and industry professionals involved in traditional 3D printing by highlighting the latest advancements in the legal and bioethical domains [8]. It will present a systematic review of scientific literature and patents to provide a cohesive overview of the existing knowledge and developments in the field [9]. This review will be particularly valuable for researchers new to the field, offering a concise summary of the companies actively involved in 3D bioprinting [10], along with a critical examination of the legal and bioethical perspectives [11].

The structure of the paper is as follows: the "Current 3D Bioprinting Market Trends" section will discuss and analyze the findings from the literature review. The subsequent section will explore the concept of 3D bioprinting technology in relation to its ethical and legal constraints, considering both educational and industrial viewpoints. Finally, the "Conclusions" section will present the key findings and insights from this study.

The Objectives of the Paper:

- To conduct a comprehensive analysis of the current trends and developments in the 3D bioprinting market through a systematic review of relevant literature.
- To examine the ethical and legal constraints of 3D bioprinting technology from the perspectives of academia and industry.
- To identify and highlight the key players and research areas in the field based on an analysis of scientific papers, patents, and companies.
- To critically evaluate the ethical and legal implications of 3D bioprinting technology and provide insights for future advancements in the field.

This paper will employ a systematic literature review methodology to gather and analyze scientific publications, patents, and company information that are relevant to the topic. The findings will be synthesized and discussed within the context of the current trends in the 3D bioprinting market. Ethical and legal considerations will be explored by examining the existing literature and incorporating viewpoints from both academia and industry. Additionally, a comparative analysis will be conducted to assess the performance and contributions of different companies and research institutions in the field of 3D bioprinting. Factors such as technological innovations, commercialization efforts, intellectual property portfolio, and research output will be taken into account.

In conclusion, this paper aims to address the existing gap in the literature by providing a comprehensive review of the legal and bioethical dimensions of 3D bioprinting technology. By exploring current market trends, ethical considerations, and key players, this study aims to contribute to the understanding of the field and offer insights for future advancements and regulatory frameworks.

The paper is structured as follows: the section on "current 3D bioprinting market trends" presents the findings from the literature review and provides a thorough analysis. The subsequent section discusses the concept of 3D bioprinting technology in relation to its ethical and legal limitations from both educational and industrial perspectives. The "Conclusions" section summarizes the key findings and implications of the study.

## 7.2 Current 3D Bioprinting Market Trends

"3D bioprinting" refers to the utilization of 3D printing techniques to combine cells, growth factors, and biomaterials, aiming to create functional components that replicate the properties of natural tissues. The overall process of 3D bioprinting typically involves layer-by-layer assembly of tissue structures using bio-inks, which find applications in diverse clinical [12] and tissue-related fields.

In the year of 2021, the market for 3D bioprinting was estimated to be worth US$ 875.33 million and is expected really to grow at a rate (Figure 7.1).

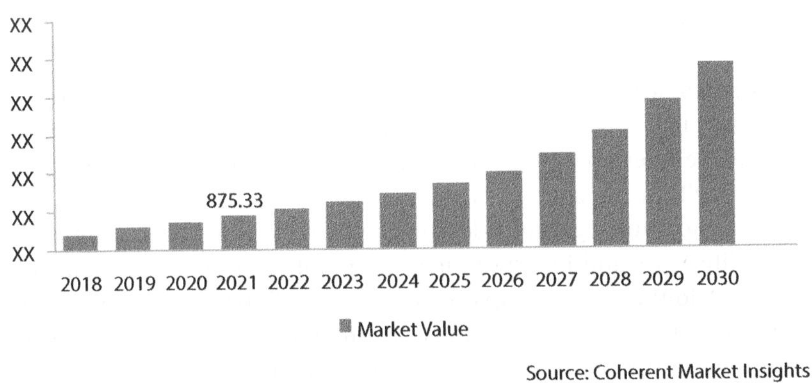


**Figure 7.1** The forecast for the global 3D bioprinting market from 2017 to 2030 in millions of dollars [13].

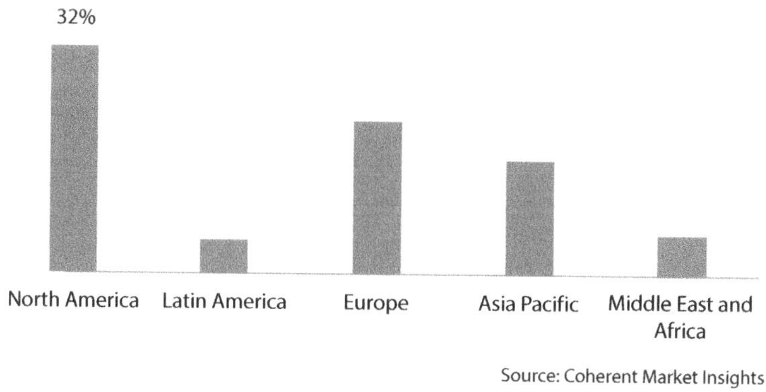


**Figure 7.2** The global market share (%) for 3D bioprinting by region in 2021 [13].

of 21 CAGR percentage between 2022 and 2030 to reach US$ 4,815.02 million. In terms of volume, North America accounted for 40% of the global 3D bioprinting market in 2021, holding a prominent position [13] (Figure 7.2).

The 3D systems embraced additive works in January 2021 to expand the AM software selection and the creation capability, which would increase with the adoption of additive production in the industrial and healthcare industries [13]. To boost the chances of using AM in clinical and high-reliability applications, 3D systems used Allevi. In a joint announcement, 3D Systems and CollPlant Biotechnologies boosted the creation in the biomedical sector. This agreement would use 3D bioprinting to promote the development of regenerative pharmaceuticals and therapies [14].

(a) Education sector 3D printing market

The integration of 3D printing technology, often referred to as additive manufacturing, has witnessed significant applications in the field of education across elementary and secondary schools, colleges, libraries, and other educational environments. This innovative concept has also had a transformative impact on the manufacturing industry [15]. Numerous manufacturers have embraced 3D printing techniques to fabricate a diverse range of products, including prosthetic limbs, aircraft components, and even pharmaceuticals. The adoption of 3D printing has revolutionized the manufacturing landscape, offering unprecedented opportunities for customization, cost-effectiveness, and streamlined production processes.

The integration of 3D printing technology in education brings about numerous benefits, fostering enhanced learning experiences for both students and teachers (Figure 7.3). By utilizing 3D printing, individuals are able

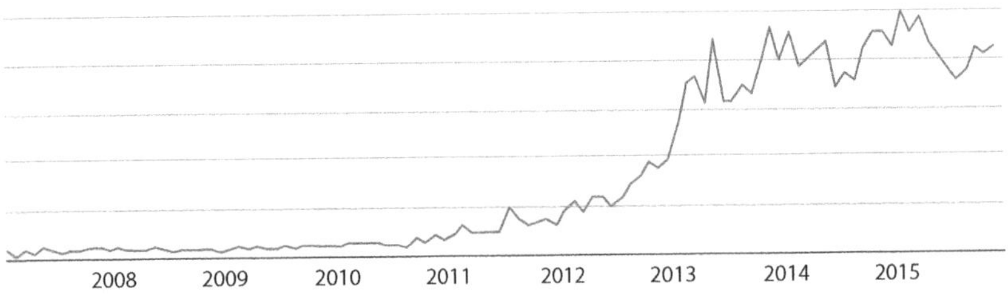


**Figure 7.3** The use of 3D printing in education [13].

to deepen their understanding of various subjects, refine their skills, and develop a heightened interest in the subject matter. Furthermore, 3D printing stimulates collaboration and nurtures creativity in the context of problem-solving [16].

Currently, companies from a variety of sectors use 3D printing goods and services to ease quick prototyping, speed up production, and streamline supply chain management. Educational institutions use 3D technology to provide a greater learning experience, much like businesses and industries do [17]. The number of educational institutions, both for-profit and nonprofit, that purchase 3D printers and use 3D printing services has been steadily rising. In the education sector, the market of 3D printing is expected to surge by USD 1.54 billion within the time period of 2020 to 2024 [18], progressing at a CAGR of over 19% over the projected period, claims a market research analysis published by Technavio [19].

A variety of three-dimensional objects are produced by educational institutions using 3D printers and CAD files. They also create three-dimensional items using different thermoplastic varieties [20]. To help pupils understand complicated ideas and concepts, teachers and instructors 3D-print a variety of items. The 3D-printed materials aid teachers in bridging the physical and digital worlds. However, the education sector has several cutting-edge 3D printing uses.

(b) 3D Printing Market in Industry area

The process of transforming a digital file into a three-dimensional object is commonly referred to as 3D printing. This technology has garnered positive responses from a multitude of industries, including manufacturing, art and jewelry, construction, aerospace, housing development, automotive, soft sensors, weapons, robotics, and computing [21]. Over the years, 3D printing has become increasingly prevalent and has revolutionized various sectors. One such field that has witnessed significant advancements in recent times is healthcare, alongside other industries. The expanding research and development efforts in 3D bioprinting have opened up vast opportunities in areas such as tissue engineering, drug delivery systems, the production of cell, tissue, and organ constructs, as well as other biomedical applications [22]. These developments have the potential to bring about transformative changes in the healthcare industry and pave the way for innovative solutions in regenerative medicine, personalized healthcare, and advanced biomedical research.

Using biomaterials and 3D printing, 3D bioprinting replicates bodily parts like natural tissues, ligaments, bones and joints, and blood vessels. Other than these, 3D bioprinting has uses in dentistry, replicating medical equipment, creating anatomical models, and researching, formulating, and developing drugs [23].

(c) Pharmaceutical and cosmetic sectors using 3D bioprinting

Pharmaceutical companies are increasingly embracing the utilization of 3D bioprinting products and technologies to facilitate the discovery and development of novel drugs. In comparison to conventional drug testing methods, 3D bioprinting offers a more cost-effective and safer approach for medication evaluation [24]. Leveraging the capabilities of 3D bioprinting, pharmaceutical companies can significantly reduce the time required for drug development, compressing the typical three to six-year timeline into a matter of hours. By employing 3D bioprinted tissues, research and development teams can conduct early-stage and preclinical trials to evaluate potential treatments. This streamlined approach not only reduces reliance on animal testing but also enhances productivity and expedites the drug discovery process [25].

The application of 3D bioprinting has gained traction, particularly within the cosmetics industry, as awareness and understanding of this technology continue to expand [26]. Furthermore, the European Commission's prohibition of animal testing for cosmetics since March 2013 has prompted numerous companies in the cosmetics and wellness sectors to adopt 3D bioprinting. This enables the development of cutting-edge 3D tissue models and innovative methods for drug testing, aligning with the ethical considerations of animal welfare and regulatory requirements. The integration of 3D bioprinting in these industries holds immense potential for advancing research, promoting product safety, and supporting the development of novel cosmetic and therapeutic solutions.

(d) Growing demand for organ transplants

The ability to produce and making of complex organs like the heart, kidneys, and lungs for transplantation, 3D printing is becoming more and more significant in the healthcare sector (Figure 7.4). Because they use cells derived from the patient's body, 3D-printed organs are less likely to present risks than the traditional replacement of failing or damaged organs, which runs the risk of the host body refusing them.

The ability to generate new organs for organ transplants and heal broken bones with 3D bioprinting holds great promise for the future of medicine. Furthermore, it enables the printing of prosthetic limbs that are able to restore patients' lost limbs. Also, 3D bioprinting permits robust development of organs and tissues than the traditional method, which requires a donor and is equally time-consuming [27]. Due to the enormous disparity between the demand for organs and the number of donors, the waiting list for transplants lengthens daily. To solve this problem, scientists are creating artificial organs. The tiny organs that experts are now working on, nevertheless, are not quite suitable for implantation into a body. Due to these unmet business needs, the development of 3D-printed transplant organs has advanced.

The global 3D bioprinting market, valued at USD 1.7 billion in 2021, is projected to experience a compound annual growth rate (CAGR) of 15.8% from 2022 to 2030 [30]. This growth is driven by factors such as an aging population with chronic respiratory conditions and a shortage of organ donors. Additionally, increased research and development investments, technological advancements, and a rising prevalence of chronic diseases are expected to positively impact the market during the forecast period (Figure 7.5). Moreover, in response to the COVID-19 pandemic, the 3D printing community has actively contributed to the production of essential medical supplies for hospitals.

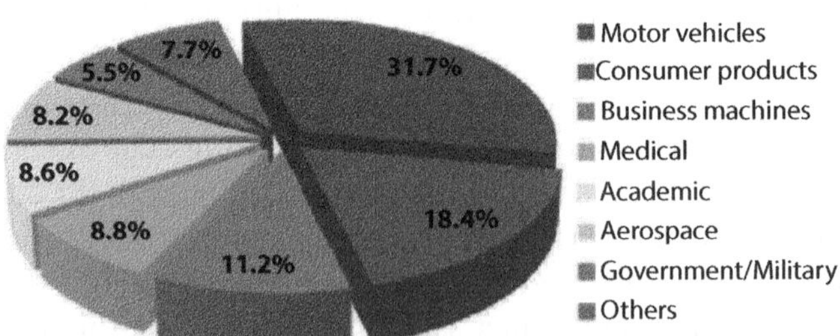


**Figure 7.4** The variety of disciplines that use 3D printing [28].

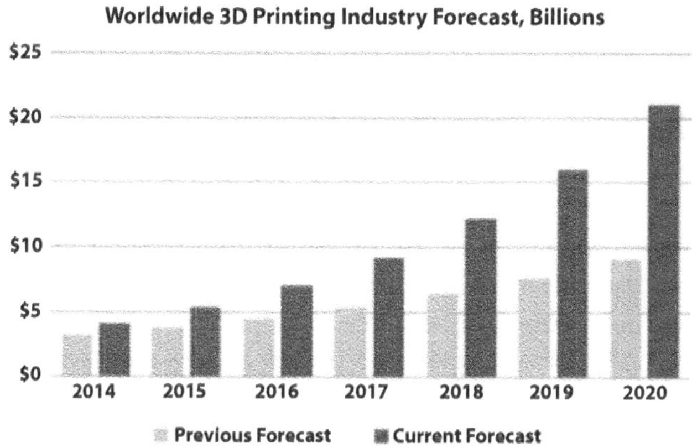


**Figure 7.5** A forecast of the global 3D printing market [29].

## 7.3 Legal and Ethical Perspectives

Bioprinting's legal considerations necessitate a broader and more focused examination, encompassing aspects such as patenting, regulations, governmental actions, and judicial decisions. One critical issue that requires immediate attention is the lack of awareness and understanding among regulatory bodies regarding this technology. As Trommelmans [31] aptly observed, addressing a few fundamental legal aspects alone will not sufficiently resolve all the complex legal concerns surrounding bioprinting, particularly within the emerging field of tissue engineering, which shares similar challenges. To draw an analogy, it would be imprudent to assume that tackling the bioethical dilemmas associated with stem cell usage and cell donation will automatically resolve all the legal intricacies entwined with this technology. Furthermore, future challenges may arise due to the absence of a dedicated legal framework specifically tailored for bioprinting. Considering the immense potential of bioprinting in the years to come, a coordinated and comprehensive legal approach is imperative to navigate the evolving landscape effectively.

(a) Problems relating to IP rights

The commercialization of emerging technologies, including bioprinting, is closely linked to patenting and intellectual property rights. Yoo [32] identifies five key areas of intellectual property (IP) in 3D bioprinting, such as materials, cell techniques, bioreactor methods, printing methods, and fabrication techniques.

However, before discussing potential patenting areas, it is crucial to determine whether bioprinting itself is patentable. Tran distinguishes between the bioprinting technique and the bioprinting product when addressing patentability. While there are existing patents related to bioprinting technology, the patentability of bioprinted goods is still unresolved.

To qualify for patent eligibility, a bioprinted product must demonstrate human ingenuity and not be a naturally occurring phenomenon. While the first criterion is met, the second criterion raises questions. Tran suggests that patents should focus on the bioprinting process rather than the products as a potential compromise. However, this may not be an ideal solution. The non-natural occurrence of bioprinted products, considering the challenge of replicating genuine tissues or organs precisely, can be a point of contention. Resolving this issue is essential to clarify ownership rights of bioprinted tissues or organs [33–37].

Intellectual property theft is another significant concern. It is projected that global losses due to 3D printing intellectual property theft could exceed $100 billion USD annually by 2018 [38]. While open-source approaches are suggested to address this issue, creators of unique bioprinted materials and devices have legitimate concerns about counterfeiting, quality control, and unlicensed products. In such cases, pursuing patent protection becomes crucial. Balancing the open-source philosophy with safety and quality in the bioprinting of complex organs requires careful consideration, particularly for organs like the brain, heart, or kidney.

However, granting patents for specific components of bioprinting technology may hinder progress in resolving challenges related to organ printing and transplantation. Therefore, it is crucial to carefully consider the implications and potential misuse of technology before deciding on a specific path regarding patenting.

(b) Rules and guidelines

The involvement of governments in research and cutting-edge technologies plays a critical role in shaping their future trajectory. Early on in technology development, policy decisions and regulatory considerations must be addressed. In the case of bioprinting, establishing a robust legal framework becomes particularly important due to its potential to revolutionize the medical industry and save lives. Smith [39] extensively studied the role of government in promoting tissue engineering and regenerative medicine, focusing on the United States.

A key regulatory concern revolves around the classification of medical products, which can fall into categories such as medical devices, medications, biologics, or "combination products" that involve a combination of devices, drugs, or biologics. This classification is significant as it determines the evaluation and approval procedures carried out by the Food and Drug Administration (FDA).

Each product classification undergoes specific regulatory reviews conducted by different agencies within the FDA, covering biologics, devices, and drugs. However, the review and approval procedures for bioprinting, which involve various components, lack precise definition. Moreover, due to the challenge of categorizing bioprinted products as devices, medications, or biologics, the classification process itself becomes complex.

Efforts have been made to identify "combination products" and assign them to specific centers within the FDA, which then determine their classification based on the principal mode of action. However, disagreements regarding classification often arise among different stakeholders, and this classification also affects subsequent marketing evaluation and approval procedures, significantly impacting the product's marketability and financial situation. Marketing drugs requires a New Drug Application (NDA), devices necessitate a pre-market application (PMA) or product development protocol (PDP) submitted by the sponsor.

Devices must obtain FDA permission through an experimental device exemption (IDE), while medicines and biologics must receive FDA approval through an investigational new drug (IND) application before proceeding to clinical approval. The intricate evaluation and approval process heavily rely on the classification of the products, leading to discussions and disputes when bioprinted products enter the market due to the complexity of the regulatory framework. Furthermore, the situation becomes even more complex on a global scale, as different countries have their own medical product regulation frameworks, as highlighted by Heinonen's investigation into the European Union's regulation of tissue-engineered products [40].

Gheisari [41] studied the advancements developed in the Islamic Republic of Iran's stem cell and tissue engineering research and emphasized the significance of governmental policies and actions, taking into account the stringent religious regulations in existence [42]. In order to tackle the governing concerns encompassing bioprinting and create a shared framework of policies, it is essential to create an international agency with representatives from various nations given the complexity and variety of cultural, social, and religious perspectives found among various populations. Such a project would enable a full assessment of the technology's acceptance, possible advantages, risks of misuse, and commercial opportunities. It would also prevent delays in the clinical deployment of this life-saving discovery establishing national regulatory organizations to investigate many areas of bioprinting, such as cultural and religious beliefs, would be a valuable starting point.

(c) Other regulatory factors
The regulatory concerns related to the classification [43], review, and approval procedures of bioprinted items are interconnected with other important factors, such as Good Manufacturing Practices (GMP) [44], which are aimed at ensuring the quality and consistency of products [45]. Upholding a safe and clean bioprinting environment, regulating the environment to prevent infection and outbreaks, creating clearly defined operations, efficient record keeping and control, abiding by proper record keeping practices, addressing and storing deviations, and preserving thorough and readily available manufacturing records, including distribution history, are just a few of the various aspects covered by GMP. GMP is implemented in several nations under a variety of titles, including "cGMP" in the United States (by the FDA), "EU-GMP" in the European Union, and the Medicines Act or "The Orange Guide" in the United Kingdom. It is critical to take the GMP tenets into account while creating the legal framework for bioprinting.

Given the significant potential of bioprinting, there is an increased risk of it being misused. Discussions should be held to assess whether licensing could help mitigate this risk. Rather than allowing unrestricted access to bioprinters, which would enable the printing of any organ by individuals who can afford the equipment (though some knowledge of cell culture and biology would be required), a more controlled approach could be considered. Licensing the technology to specific medical facilities or qualified individuals could be explored. While opinions may vary on this topic, it is important to include it in the discussions concerning potential regulations governing bioprinting technology.

## 7.4 Regarding the Introduction and Advancement of 3D Bioprinting

### 7.4.1 Current and Potential Paths for Bioethical Discourse

The development of bioprinting technologies sparks profound discussions on various aspects of human nature, including projects related to human enhancement, concerns regarding technological design, youth extension, and the possibility of technological immortality [44]. Rejuvenation technologies that aim to reverse aging and consciousness transfer technologies fall into the category of technological immortality programs, exploring the transfer of human personality. Bioprinting new organs and replacing aging ones fall under rejuvenation technologies, prompting a consideration of the indications and limitations of bioprinting technologies. The practicality and moral implications of creating these technologies pose initial challenges associated with bioprinting organs and tissues. Although we have the capability to print organs, the crucial question remains: why and for what purpose? Considering the demand for organ transplantation and the global shortage of organ donations, the solution seems clear.

Technological advancements have the potential to save numerous lives. Most governments and international organizations currently view technology as morally acceptable when it has a positive therapeutic impact [46]. Utilitarian philosophy provides a fundamental framework for supporting the broad application of biotechnology, as stated by G. B. Yudin. Various perspectives exist, ranging from radical transhumanism that places no restrictions on human technological development (human enhancement) to more conservative variants that acknowledge the societal risks associated with human transformation. However, given that human modification programs are driven by individuals' desire for self-improvement and a better life, the utilitarian framework of ethical reasoning tends to provide greater support.

As medical biotechnologies expand and become more complex, it becomes challenging to differentiate between therapy and human betterment [47]. Before therapeutic 3D bioprinting becomes widely used in patient care, it is crucial to monitor the pace of technological development and investigate ethical issues (Figure 7.6). Due to the use of living cells in the bioprinting process, 3D bioprinting requires special consideration due to its unique ethical and legal aspects. The "cell" stands as a crucial element in bioprinting, significantly influencing the properties of the bioprinted tissue.

Ethical concerns related to donation arise with allogeneic cell transplantation, including donor anonymity, informed donor consent, potentially invasive cell generation processes, and ownership of donor cells. Another of the key cornerstones of medical ethics, the maxim "primum non nocere" (first, do no harm), is still up for discussion among experts and professionals [48]. The misuse principle takes into account the morality of the action, the agent's aim, the method of execution, and the balance between favorable and unfavorable results and any potential bad effects.

Stem cells often serve as the fundamental components for biofabricating human tissues and organs. Ethical concerns primarily revolve around the source of these stem cells. The use of human embryonic stem cells (ESCs) has faced significant criticism and is subject to appropriate legal and moral restrictions. Obtaining ESCs involves ethical considerations regarding the moral status of embryos, the legality of terminating pregnancies, and human participation in research. In addition to the legal and ethical

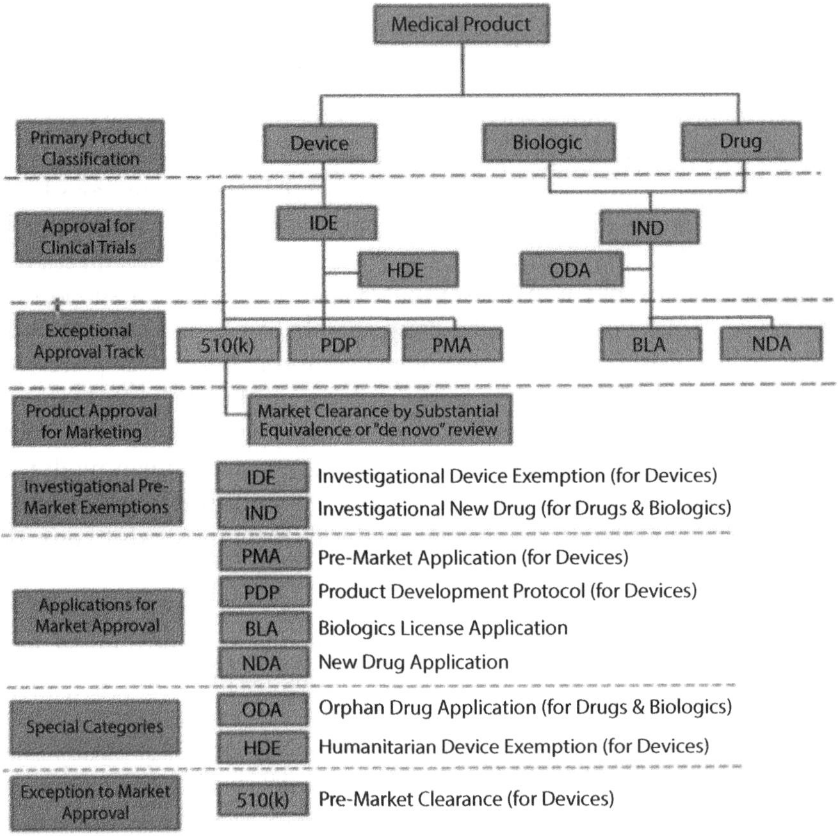


**Figure 7.6** Regarding the introduction and advancement of 3D-bioprinting, current and potential paths for bioethical discourse [39].

concerns surrounding ESCs, the use of bio-inks derived from allogeneic cells may raise ethical considerations. It is crucial to evaluate the issues associated with using stem cells obtained from coerced or uninformed consent donors under duress. Furthermore, commercialization poses challenges, as certain legal regulations prohibit the industrial and commercial use of ESC-based 3D printing technology, considering human embryos as ineligible for patent rights [49, 50].

In this context, the social and theological implications of utilizing animal cells must be taken into account. Patients undergoing xenotransplantation may face psychosocial challenges related to their identity and self-perception. Moreover, individuals practicing specific religions may have objections to the use of cells derived from certain animal species [51, 52]. To address these moral concerns, induced pluripotent stem cells (iPSCs) can be generated by reprogramming differentiated cells [53], eliminating the need for xenogeneic or embryonic stem cells [54]. iPSCs have the ability to differentiate into various types of adult body cells [55]. However, it is important to note that utilizing autologous iPSCs as bio-ink for 3D printing human organs is not morally neutral, despite using the patient's own cells [56, 57]. This approach still has limitations, and cell reprogramming technology is still not fully developed.

Prior to transplantation, it has become very important to guarantee appropriate differentiation of all stem cells [58]. The possibility of tumorigenicity is a major worry when employing iPSCs [59]. To guarantee the security of iPSC-based therapeutics, stem cell lines intended for clinical applications must undergo genetic analysis. Nevertheless, the gathering, keeping, and application of human genetic data bring up fresh moral and legal concerns. Increased access to an individual's genomic information for research purposes raises the risk of compromise or malicious use [60]. The "Privacy and Progress in Whole Genome Sequencing" report, issued by The Presidential Commission for the Study of Bioethical Issues in the United States in 2012, provides recommendations for preserving individual privacy while enabling the exchange of genetic data [61].

In Russia, there is a lack of specific legislation governing human genome research, genetic companies, and the handling of genetic data [62]. The absence of legal definitions and requirements for processing and transmitting genetic data as a distinct category of personal data, as well as obtaining consent from donors for research participation, is a notable gap [63]. Current laws do not adequately protect donors' rights, place restrictions on the use of biological materials obtained from donors for scientific research, or provide clear guidelines for obtaining ethical committee approval for 3D bioprinting research. Additionally, the ethical considerations of bioprinting, which relies on digital models, need to be taken into account. The advancement of 3D bioprinting blurs the boundaries between the physical and digital realms, resulting in the "digitalization" of real-world objects and the digitization of individuals. This raises important questions about the responsibility for creating and evaluating these digital models. Who should be responsible for translating anatomical images into digital form—designers, biologists, or programmers—and to what extent? What are the legal rights of the model's owner? Can the model be utilized without the patient's consent? Are commercial uses of these models permissible [64]? The digitization of humans also brings concerns about privacy and confidentiality, as the digital 3D model represents personalized human data within the context of bioprinting technology.

To ensure the protection of privacy, it is crucial to establish specific regulations governing the collection, storage, handling, processing, and usage of such information [65]. In the realm of biomedical research, informed consent and confidentiality play vital roles in safeguarding human rights. Medical confidentiality, closely linked to the principle of confidentiality, seeks to protect patients' lives while preserving the privacy of treatment details and personal characteristics. Establishing trust-based relationships is essential for delivering timely and effective medical care [66]. Informed consent is a fundamental principle deeply rooted in the ethical and legal framework of medical practice. It is widely recognized and referenced in various international and national legal documents. The Nuremberg Code, for instance, ensures comprehensive informed consent for individuals participating in medical experiments, encompassing a clear understanding of the nature, duration, purpose, procedures, and potential risks involved. Similarly, the Convention on Human Rights and Biomedicine emphasizes that medical interventions should only be performed with voluntary written consent from individuals who have received adequate information about the intervention's purpose, nature, consequences, and risks.

Obtaining informed consent for 3D bioprinting can be challenging in emergency cases where patients are unable to provide consent. Additionally, situations involving participants lacking decision-making capacity or comprehension of donation choices, such as patients in intensive care units, can complicate the process of obtaining informed permission. Conducting clinical trials is essential for integrating 3D bioprinting technology into clinical practice and addressing associated risks.

The planning of human clinical trials for 3D bioprinting of human tissues and organs presents unique challenges in addition to ethical guidelines and requirements for human participation in treatments. As personalized medicine drives the advancement of 3D bioprinting technology, each biofabricated product is specifically designed for an individual, necessitating modifications to the experimental design for each case. Traditional approaches to clinical trials, such as double-blind, randomized control studies, cannot be directly applied to 3D bioprinting due to the uniqueness of each treatment tailored to an individual's specific symptoms. Extrapolating the outcomes of one case to subsequent therapies is limited.

Nevertheless, standardized criteria and methodologies can be established based on initial clinical studies, even when the biomaterials used are customized. However, organizing experimental investigations on 3D bioprinting of human organs poses significant challenges. The study must adhere to ethical principles and prioritize participant safety. Each patient becomes the initial subject as the effectiveness and safety of custom-made organs cannot be tested on other individuals. Factors such as the risks-to-benefits ratio and inclusion criteria, including the involvement of end-of-life patients in research, become important considerations in this context.

Another concern in clinical investigations of therapies assisted by 3D bioprinting is the issue of participant withdrawal from the study. Individuals enrolled in a 3D printing study may have trouble invoking the right to end the trial following the implantation of an artificial organ, in contrast to conventional clinical trials where drug dosage can be gradually reduced. Reversing interventions in 3D bioprinting therapies may have limitations in terms of process reversibility, and attempting to remove the implant and any proliferated cells may introduce additional risks to patients. Most importantly, participation in a bioprinting trial may exclude patients from receiving alternative treatments.

### 7.4.2 Legal Concerns with the Introduction of 3D Bioprinting Into Clinical Practice

It is a matter of fact that there is a prevalence of notable risks and unpredictable conditions linked with technology with regard to legal aspects and regulation of 3D Bioprinting as there is no worldwide co-operation for an effective mechanism to handle the same. The involvement of multiple stakeholders in the bioprinting production process further complicates the issue. In order to extract results for the development and unification of both fields i.e., Bioprinting and the Clinical practice, collaborative efforts is needed from the experts from the fields such as ethical, law, designing, engineering, Biology, Medical etc. While a comprehensive regulatory framework for overseeing the entire bioprinting process is currently lacking, and also the fields of regenerative medicine and tissue engineering has already been put with regulation [67].

The therapies of gene and somatic cells and its related tissue-engineered products have been categorized as Therapy Medical products by the European Medicines Agency [68]. The concepts such as controlling significant hazards and setting up of duty for ensuring quality of the products is much important. Determining who should be responsible for quality control and who should bear liability if the recipient of a bioprinted organ raises quality-related concerns are key aspects to address. In Russia, significant attention has been given to the legal issues surrounding the production and use of bioprinted human organs, and discussions on potential regulatory frameworks for bioprinting technology have taken place [69]. Currently, work contracts or contracts for medical services can be used to control the relationship between 3D bioprinting service providers, patients, and healthcare organizations. These agreements are suitable when custom-made organs or tissues are created based on the customer's specifications [70, 71].

The commercialization of 3D bioprinted human organs raises bioethical and legal concerns due to the significant differences between the final product and its original biomaterial source. However, it may be deemed acceptable for experts in the field of bioprinting to engage in commercializing these products. Bioprinted organs have distinct objective characteristics that differentiate them from natural human organs. Since artificial organs are produced through regulated and intentional biofabrication processes outside the human body, they do not pose significant risks to the life and health of the cell donor.

Moreover, informed consent primarily applies to the creation of biological material rather than the biofabrication of the organ itself. Before being transplanted, 3D bioprinted organs may have implications for civil rights due to their unique characteristics. Consequently, it is reasonable to expect limited commercialization of biofabricated organs and tissues. When these tissues and organs have been put in, it is important to acknowledge them as essential parts of the body of a person. It is important to recognize the legal importance of organs as objects, acknowledging the loss of control and the related constraints on their removal, especially in light of the current prohibitions and limitations on the removal and purchase of human biological material.

3D bioprinting operations are legally governed by a number of intellectual property rights, including patents, copyrights, design rights, and trademarks. State participation in the study and development of novel innovations is essential as it will influence the technology's growth course. Political and normative issues must be addressed at the early stages of technological development. The advancement of 3D bioprinting technology has the potential to revolutionize the medical industry and save lives, thus it is important to give it considerable thought and call for the creation of the necessary legal framework [72].

Introduction:

Advancements in biotechnology have led to ground breaking developments in the field of 3D bioprinting, allowing researchers to create three-dimensional structures that mimic human tissues and organs. While this technology holds immense potential for medical applications such as organ transplantation and drug testing, it also raises several legal and ethical concerns. This case study explores a real-world scenario involving 3D bioprinting and delves into the legal and ethical issues surrounding its use.

For our better understanding, it is important to look into a case study. Where, a renowned research institute has made significant progress in 3D bioprinting, specifically in the field of regenerative medicine. The researchers have successfully developed a technique to bioprint functional human liver tissue using patient-specific cells. This breakthrough has the potential to revolutionize organ transplantation by eliminating the need for donors and reducing the risk of rejection. However, as the research progresses, legal and ethical concerns begin to emerge.

Legal Issues involved in this case was with regard to Intellectual Property Rights, Regulatory approval, and Liability and Malpractice. The research institute faces challenges regarding the ownership and protection of their bioprinting techniques, processes, and biomaterial formulations. Competing organizations or individuals may attempt to claim ownership or develop similar technologies, potentially leading to disputes and patent infringement claims. 3D bioprinted organs and tissues are subject to regulatory approval, just like traditional transplantation procedures. The research institute must navigate the complex regulatory landscape to ensure compliance with safety and efficacy standards. Issues such as product labeling, quality control, and ethical considerations related to the use of human cells and tissues may arise. If 3D bioprinted organs or tissues are used in transplantation procedures, questions regarding liability in case of adverse effects or malpractice may arise. Determining responsibility between the research institute, medical professionals, and regulatory bodies can be legally challenging and may require new frameworks and precedents.

Ethical Issues involved in the case was informed consent, Equity and Access and Privacy and Genetic information. Obtaining informed consent from patients or donors for the use of their cells or tissues in 3D bioprinting is critical. Ethical concerns arise regarding the level of understanding and disclosure of potential risks and uncertainties associated with this emerging technology. The cost and availability of 3D bioprinting technology may create disparities in access to life-saving treatments. Ensuring equitable distribution and affordable access to 3D bioprinted organs and tissues present ethical challenges, as wealthier individuals or regions may have an advantage over others. The process of acquiring patient-specific cells for bioprinting involves collecting and analyzing genetic information. Safeguarding this sensitive data raises concerns about privacy, genetic discrimination, and potential misuse of genetic information by third parties.

As 3D bioprinting technology continues to advance, legal and ethical issues surrounding its use become increasingly significant. This case study highlights some of the key concerns related to intellectual property rights, regulatory approval, liability, informed consent, equity, access, and privacy. Addressing these issues will require collaborative efforts among researchers, policymakers, legal experts, and ethicists to strike a balance between innovation and protecting the rights and well-being of individuals and society as a whole.

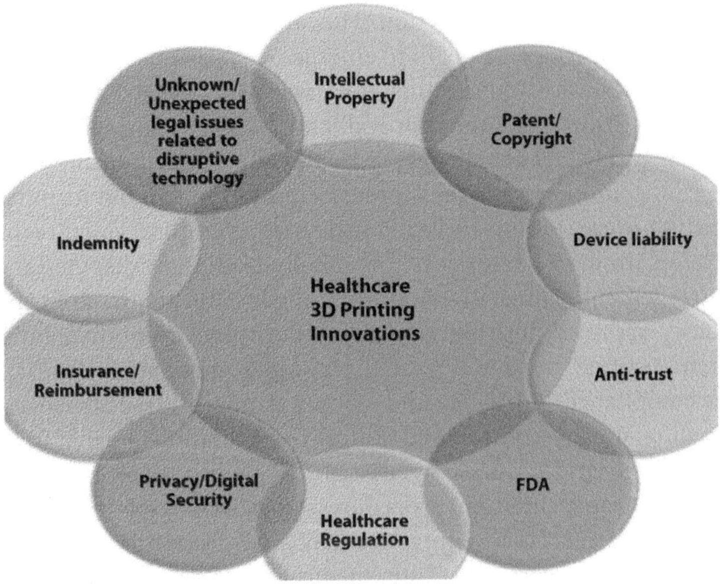


### 7.4.3 Ethical Concerns with the 3D Bioprinting of Artificial Ovaries and Their Use in Clinical Settings

The biofabrication of reproductive organs, such as ovaries, testicles, and uteruses, serves a different purpose compared to organ bioprinting as its primary focus is not on life-threatening conditions but rather on enhancing quality of life. In women of reproductive age, gonadotoxic oncological therapies can result in primary ovarian insufficiency. To preserve fertility, various techniques like cryopreservation of ovarian tissue and oocytes have been developed [73].

There are presently no treatments available to help patients with ovarian cancer or cancers that damage the ovaries (such leukemia, neuroblastoma, and Burkitt lymphoma) become fertile again after recovery. In these people, ovarian endocrine and reproductive function are permanently impaired as a result of ovariectomy or gonadotoxic oncological therapies. For this specific group of patients, artificial ovary technology offers promise for fertility preservation and the ability to create genetically related offspring.

Encouraging results have been observed in animal trials, and ongoing advancements in this technology are taking place. However, before human clinical trials can be initiated, it is essential to address significant ethical and legal challenges associated with 3D bioprinting of artificial ovaries at this stage of technological development. The aim of the bioprinted ovary project is to tackle the complexities of human reproduction along with the ethical and legal concerns previously discussed.

As a result, a thorough evaluation of potential risks is essential at every stage of the technology, spanning from bioink production to organ transplantation. Artificial ovary technology carries risks not only for the woman but also for her future offspring, which is a critical consideration. The main concern is that manipulations of ovarian follicles outside the body could lead to genetic and epigenetic abnormalities in egg cells, adversely impacting the health of the offspring. Additionally, inherited diseases are often associated with cancer, raising the possibility of transferring cancer-related genetic variations to future generations through the development of artificial ovaries, thereby artificially increasing the number of individuals affected by cancer.

The introduction of artificial ovaries through 3D printing technology raises questions about the right to know for children conceived through this method. Could the knowledge of being conceived using artificial ovaries cause distress or potential stigmatization in society? Another significant aspect of ovary bioprinting technology is the considerable time gap between ovary removal and the potential for artificial ovary transplantation following oncological treatment.

When an organ is printed, a patient may choose not to undergo transplantation, but the high cost of the technology can create emotional pressure and a desire to become a parent. The voluntary rejection of transplantation after an artificial organ has been printed raises concerns about the morality and legality of organ ownership.

The commercialization of the technology and the resulting ethical dilemma are also significant issues associated with the development of bioprinted ovaries. While Russia's legal system allows business transactions related to assisted reproductive technology, such as paid *in vitro* fertilization (IVF) services in clinics, commercial relationships in transplantation are prohibited by Russian law. This ban on organ sales aligns with the fundamental principle of moral interactions between individuals. The scarcity of transplantable organs is one of the main driving factors behind the current trend towards commercialization. While 3D printing technology can address the issue of organ shortage, there will be challenges in terms of technology accessibility for all individuals.

It is evident that only the affluent will be able to afford the benefits of 3D bioprinting, leading to a widening social divide between the wealthy and the impoverished. The expensive nature of this scientific and technical solution raises ethical concerns regarding the social stratification caused by bioprinting, as only a select few from a specific group may have access to the commercial fabrication and transplantation of artificial ovaries [74].

Concerning the requirements for using synthetic ovaries, several questions are raised. Can this technology be available to people of all ages? Will people who already had children be able to have synthetic ovary transplants? Women with artificial ovaries can presently engage in IVF treatments, but how might having an artificial ovary effect a patient's physical and mental health? How will the end goal of attaining live birth be held accountable if it is not achieved? Programmers, engineers, biologists, and doctors. Should low-income and disadvantaged persons have to pay for the creation and transplantation of artificial ovaries?

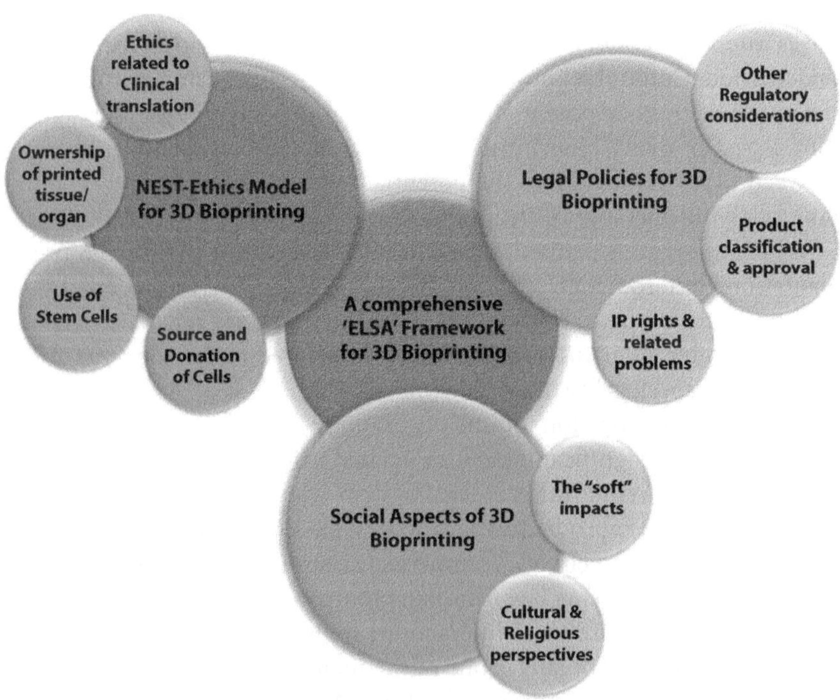

## 7.5 Conclusion

The field of 3D bioprinting raises numerous bioethical and legal concerns, and currently, there are more questions than answers. The rapid advancement of knowledge and technology has outpaced our understanding of the ethical and legal implications associated with this growth. At present, there is a lack of appropriate legal frameworks and specific regulatory requirements for the development and use of 3D bioprinted tissues and organs, not only in Russia but globally. The use of 3D bioprinting in organ transplants presents significant ethical and legal dilemmas.

The challenge in assessing the ethics of 3D bioprinting and implementing legal regulations lies in the fact that conventional clinical studies and existing regulatory standards do not fully capture the complexities of this technology. However, before widespread therapeutic use of 3D bioprinting can occur, several guidelines for scientific research and medical practice must be established. This includes developing informed consent protocols for material donation, manipulation, storage, and subsequent use, considering both commercial and research purposes. Additionally, standards for the efficacy, quality, and safety of 3D bioprinting processes and the resulting products must be defined, while also upholding the rights and dignity of individuals involved.

The formation of committees responsible for establishing and regulating national guidelines on technical, legal, and ethical aspects of 3D bioprinting technology is of utmost importance. Legal protections should be in place to safeguard all patients, including children and those lacking decision-making capacity. It is essential to set restrictions on the commercialization of 3D bioprinting technologies for human organs and tissues, manage turnover of those technologies, and create sanctions for the unlawful trafficking of artificial parts. To ensure the responsible and moral growth and utilization of 3D bioprinting technology, several actions are required.

## 7.6 Future Scope

The future holds immense potential for the scope of 3D bioprinting, both in terms of technological advancements and its impact on various industries. Here are some potential developments and opportunities that lie ahead:

Advancements in Tissue and Organ Engineering: Continued research and development in 3D bioprinting technology will lead to significant advancements in tissue and organ engineering. We can expect the production of more complex and functional organs, such as hearts, lungs, and kidneys, which could revolutionize transplantation and address the global organ shortage crisis. This could potentially save countless lives and improve the quality of life for patients with organ failure.

Personalized Medicine and Patient-Specific Treatments: 3D bioprinting enables the creation of patient-specific tissues and organs, tailored to individual needs. This opens up avenues for personalized medicine, where treatments and therapies can be designed based on a patient's specific anatomy and genetic profile. Customized implants, prosthetics, and drug delivery systems can also be developed, leading to more effective and targeted treatments.

Pharmaceutical Testing and Drug Development: 3D bioprinting can facilitate the development of realistic tissue models for pharmaceutical testing, replacing traditional animal models and improving the accuracy of preclinical trials. This technology allows for the creation of organ-on-a-chip systems, which mimic the functionality and response of human organs, enabling more efficient and reliable drug development processes.

Biofabrication in Regenerative Medicine: 3D bioprinting can play a crucial role in regenerative medicine by enabling the fabrication of patient-specific tissues and bioactive scaffolds that promote tissue regeneration. This could revolutionize the treatment of various conditions, including bone and cartilage defects, skin injuries, and nerve damage. Biofabricated tissues and scaffolds can be implanted to stimulate the body's natural healing processes and restore normal tissue function.

Integration with Robotics and AI: The integration of 3D bioprinting with robotics and artificial intelligence (AI) holds immense potential. Robotic systems can assist in the precise placement of bioink and cells, enhancing the accuracy and speed of the bioprinting process. AI algorithms can optimize the design and fabrication process, improving the quality and functionality of the printed tissues and organs.

Ethical and Regulatory Considerations: As 3D bioprinting continues to advance, it will be crucial to address the ethical and regulatory challenges associated with this technology. Clear guidelines and regulations need to be established to ensure the responsible and ethical use of 3D bioprinting, addressing concerns such as patient safety, privacy, informed consent, and equitable access to these technologies.

In conclusion, the future of 3D bioprinting holds great promise in revolutionizing healthcare, personalized medicine, and regenerative therapies. With further advancements and responsible implementation, this technology has the potential to transform the way we treat diseases, repair and replace damaged tissues and organs, and improve overall patient outcomes.

## References


1. Ng, W.L., Chua, C.K., Shen, Y.F., Print me an organ! Why we are not there yet. *Prog. Polym. Sci.*, 97, 101145, 2019. https://doi.org/10. 1016/j.progpolymsci.2019.101145
2. Choudhury, D., Anand, S., Naing, M.W., The arrival of commercial bioprinters—Towards 3D bioprinting revolution! *Int. J. Bioprint.*, 4, 2, 3, 2018. https://doi.org/10.18063/IJB.v4i2.139.
3. Hölzl, K., Lin, S., Tytgat, L. *et al.*, Bioink properties before, during and after 3D bioprinting. *Biofabrication*, 8, 3, 032002, 2016. https://doi.org/10.1088/1758-5090/8/3/032002.
4. Hospodiuk, M., Dey, M., Sosnoski, D. *et al.*, The bioink: A comprehensive review on bioprintable materials. *Biotechnol. Adv.*, 35, 2, 217–239, 2017. https://doi.org/10.1016/j.biotechadv.2016.12.006.
5. Murphy, S.V. and Atala, A., 3D bioprinting of tissues and organs. *Nat. Biotechnol.*, 32, 8, 773–785, 2014. https://doi.org/10.1038/nbt.2958.
6. Li, J., Chen, M., Fan, X. *et al.*, Recent advances in bioprinting techniques: Approaches, applications and future prospects. *J. Transl. Med.*, 14, 1, 271, 2016. https://doi.org/10.1186/ s12967-016-1028-0.
7. Ozbolat, I.T., Moncal, K.K., Gudapati, H., Evaluation of bioprinter technologies. *Addit. Manuf.*, 13, 179–200, 2017. https://doi.org/ 10.1016/j.addma.2016.10.003.
8. Rodríguez-Salvador, M., Rio-Belver, R.M., Garechana-Anacabe, G., Scientometric and patentometric analyses to determine the knowledge landscape in innovative technologies: The case of 3D bioprinting. *PloS One*, 12, 6, 5, 2017. https://doi.org/10.1371/journal.pone.0180375.
9. Agarwala, S., Lee, J.M., Ng, W.L. *et al.*, A novel 3D bioprinted fexible and biocompatible hydrogel bioelectronic platform. *Biosens. Bioelectron.*, 102, 365–371, 2018. https://doi.org/10.1016/j.bios. 2017.11.039.
10. Munoz-Abraham, A.S., Rodriguez-Davalos, M.I., Bertacco, A. *et al.*, 3D printing of organs for transplantation: Where are we and where are we heading? *Curr. Transpl. Rep.*, 3, 1, 93–99, 2016. https:// doi.org/10.1007/s40472-016-0089.
11. Antoshin, A.A., Churbanov, S.N., Minaev, N.V. *et al.*, LIFTbioprinting, is it worth it? *Bioprinting*, 15, e00052, May 2019. https:// doi.org/10.1016/j.bprint.2019.e00052.
12. Jin, Y., Ji, S., Li, X. *et al.*, A scientometric review of hotspots and emerging trends in additive manufacturing. *J. Manuf. Technol. Manage.*, 28, 1, 18–38, 2017. https://doi.org/10.1108/JMTM-12-2015-0114.

13. 3D Bioprinting Market, Size, Trends and Forecast to 2030, 2023, https://www.coherentmarketinsights.com/market-insight/3d-bioprinting-market-5166.
14. Ramos, T. and Moroni, L., Tissue engineering and regenerative medicine 2019: The role of biofabrication—A year in review. *Tissue Eng. Part C Methods*, 26, 2, 91–106, 2020. https://doi.org/10.1089/ ten.tec.2019.0344.
15. Moroni, L., Boland, T., Burdick, J.A. *et al.*, Biofabrication: a guide to technology and terminology. *Trends Biotechnol.*, 36, 4, 384–402, 2018. https://doi.org/10.1016/j.tibtech.2017.10.015.
16. Arshad, J., Helms, R.W., Orge, F.H., Sayegh, R.R., Educational Value of a 3D Printer in Ophthalmology Training. *J. Acad. Ophthalmol.*, 10, 1, e69–e7, 2018.
17. Augusto, I., Monteiro, D., Girard-Dias, W., dos Santos, T.O., Rosa Belmonte, S.L., Pinto de Oliveira, J., Guimarães, M.C.C., Virtual reconstruction and three-dimensional printing of blood cells as a tool in cell biology education. *PloS One*, 11, 8, e0161184, 2016.
18. Becker, S.A., Brown, M., Dahlstrom, E., Davis, A., DePaul, K., Diaz, V., Pomerantz, J., *NMC Horizon Report: 2018 Higher Education Edition*, EDUCAUSE, Louisville, CO, 2018.
19. Bernhard, J.-C., Isotani, S., Matsugasumi, T., Duddalwar, V., Hung, A.J., Suer, E., Gill, I.S., Personalized 3D printed model of kidney and tumor anatomy: A useful tool for patient education. *World J. Urol.*, 34, 3, 337–345, 2016.
20. Canessa, E., Fonda, C., Zennaro, M., *Low-cost 3D Printing for Science, Education and Sustainable Development*, The Abdus Salam International Centre for Theoretical Physics (ICTP), Trieste, Italy, 2013.
21. Chytas, C., Diethelm, I., Tsilingiris, A., Learning programming through design: An analysis of parametric design projects in digital fabrication labs and an online makerspace. *2018 IEEE Global Engineering Education Conference (EDUCON)*, April 17–20, 2018, 2018.
22. Li, J. *et al.*, Recent advances in bioprinting techniques: Approaches, applications and future prospects. *J. Transl. Med.*, 14, 271, 2016.
23. Angelopoulos, I. *et al.*, Engineering inkjet bioprinting processes toward translational therapies. *Biotechnol. Bioeng.*, 117, 1, 272–284, 2020.
24. Kilmer, S., iBio inks 3D bioprinting agreement. *GlobeNewswire*, published online, 2019. https://www.globenewswire.com/news-release/2019/08/19/1903773/0/en/iBio-Inks-3D-Bioprinting-Agreement.html.
25. Lee, A. *et al.*, 3D bioprinting of collagen to rebuild components of the human heart. *Science*, 365, 6452, 482–487, 2019.
26. Skylar-Scott, M.A. *et al.*, Biomanufacturing of organ-specific tissues with high cellular density and embedded vascular channels. *Sci. Adv.*, 5, 9, eaaw2459, 2019.
27. Grigoryan, B. *et al.*, Multivascular networks and functional intravascular topologies within biocompatible hydrogels. *Science*, 364, 6439, 458–464, 2019.
28. Https:/natgeoeducationblog.files.wordpress.com/2013/11/rapidprototyping worldwide by zureks.png

29. Columbus, L., Roundup of 3D printing market forecasts and estimates, 2015. http://www.forbes.com/sites/louiscolumbus/2015/03/31/2015-roundup-of-3d-printing-market-forecasts-and-estimates/#5d0a7e6c1dc6.
30. Costa, P.F., Translating biofabrication to the market. *Trends Biotechnol.*, 37, 10, 1032–1036, 2019.
31. Trommelmans, L., Selling, J., Dierickx, K., Is tissue engineering a new paradigm in medicine? Consequences for the ethical evaluation of tissue engineering research. *Med. Health Care Philos.*, 12, 459–467, 2009.
32. Yoo, S.S., 3d-printed biological organs: medical potential and patenting opportunity. *Expert Opin. Ther. Pat.*, 25, 507–511, 2015. [91] Tran, J.L., Patenting bioprinting. *Harvard J. Law Technol. Digest*, 2015.
33. Resnik, D.B., The commercialization of human stem cells: Ethical and policy issues. *Health Care Anal.*, 10, 127–154, 2002.
34. Dhai, A. *et al.*, Ethical and legal controversies in cloning for biomedical research-a South African perspective. *S. Afr. Med. J.*, 94, 906, 2004.
35. Salvaterra, E. *et al.*, The ethics of cord blood banking in light of ownership, informed consent, and solidarity. *Cell Preserv. Technol.*, 4, 91–96, 2006.
36. Samanta, A., Samanta, J.O., Price, D., Who owns my body–thee or me? *Hum. Tissue story continues. Clin. Med.*, 4, 327–331, 2004.
37. Thasler, W.E. *et al.*, Charitable state-controlled foundation human tissue and cell research: Ethic and legal aspects in the supply of surgically removed human tissue for research in the academic and commercial sector in Germany. *Cell Tissue Bank.*, 4, 49–56, 2003.
38. Wolinsky, H., Printing organs cell-by-cell. *EMBO Rep.*, 15, 8, 836–838, 2014.
39. Smith, D.S., The government's role in advancing regenerative medicine and tissue engineering–science, safety, and ethics. *Periodontology*, 41, 16–29, 2006.
40. Heinonen, M., Oila, O., Nordström, K., Current issues in the regulation of human tissue-engineering products in the European Union. *Tissue Eng.*, 11, 1905–1911, 2005.
41. Gheisari, Y. *et al.*, Stem cell and tissue engineering research in the Islamic republic of Iran. *Stem Cell Rev. Rep.*, 8, 629–639, 2012.
42. Yudin, B.G., Institutionalization of bioethics and human enhancement, in: *Workbooks Bioethics*, vol. 23, pp. 5–11, 2016.
43. Tishchenko, P.D., Designing of the person: Ideals and technologies, in: *Workbooks Bioethics*, vol. 24, pp. 36–64, 2015.
44. Bushev, S., The philosophical and natural-science aspects of the study of the problem of aging and technological immortality. *Soc Poli. Nauki.*, 3, 198–200, 2018.
45. Etzioni, A., On a communitarian approach to bioethics. *Theor. Med. Bioeth.*, 32, 5, 363–374, 2011. https://doi.org/10.1007/s11017-011-9187-8

46. Yudin, B., Technoscience and "human enhancement". *Epistemol. Philos. Sci.*, 48, 2, 18–27, 2016.
47. Gilbert, F., O'Connell, C.D., Mladenovska, T. *et al.*, Print me an organ? Ethical and regulatory issues emerging from 3D bioprinting in medicine. *Sci. Eng. Ethics*, 24, 1, 73–91, 2018.
48. Gilbert, F., Viaña, J.N.M., O'Connell, C.D. *et al.*, Enthusiastic portrayal of 3D bioprinting in the media: Ethical side effects. *Bioethics*, 32, 2, 94–102, 2018.
49. Directive 2004/23/EC of the European parliament and of the council of 31 March 2004 on setting standards of quality and safety for the donation, procurement, testing, processing, preservation, storage and distribution of human tissues and cells 2004. *Off. J. Eur. Union*, 102, 48–58, 2004.
50. Russian Federation, Federal Law dated 23.06.2016 No. 180- FZ, in: *Biomedical Cellular Products*, Russian Federation, Moscow, 2016.
51. Lo, B. and Parham, L., Ethical issues in stem cell research. *Endocr. Rev.*, 30, 204–213, 2009.
52. Volarevic, V., Markovic, B.S., Gazdic, M. *et al.*, Ethical and safety issues of stem cell-based therapy. *Int. J. Med. Sci.*, 15, 36–45, 2018.
53. Gulyaev, V.A., Khubutiya, M.S., Novruzbekov, M.S. *et al.*, Xenotransplantation: History, problems and development prospects. *Transplantologiya Russ. J. Transplant.*, 11, 37–54, 2019.
54. Blum, B. and Benvenisty, N., The tumorigenicity of diploid and aneuploid human pluripotent stem cells. *Cell Cycle*, 8, 3822–3830, 2009.
55. Nori, S., Okada, Y., Nishimura, S. *et al.*, Long-term safety issues of iPSC-based cell therapy in a spinal cord injury model: Oncogenic transformation with epithelialmesenchymal transition. *Stem Cell Rep.*, 4, 360–373, 2015.
56. Tang, W., Challenges and advances in stem cell therapy. *Biosci. Trends*, 13, 286–286, 2019.
57. Andrews, P.W., Ben-David, U., Benvenisty, N. *et al.*, Assessing the safety of human pluripotent stem cells and their derivatives for clinical applications. *Stem Cell Rep.*, 9, 1–4, 2017.
58. Sullivan, S., Stacey, G.N., Akazawa, C. *et al.*, Quality control guidelines for clinical-grade human induced pluripotent stem cell lines. *Regen. Med.*, 13, 859–866, 2018.
59. Jo, H.Y., Han, H.W., Jung, I. *et al.*, Development of genetic quality tests for good manufacturing practice-compliant induced pluripotent stem cells and their derivatives. *Sci. Rep.*, 10, 1, 3939, 2020.
60. The Presidential Commission for the Study of Bioethical Issues, *Privacy and Progress in Whole Genome Sequencing*, 2012, Available at SSRN: https://ssrn.com/abstract=2459932
61. Vasiliev, S.A., Osavelyuk, A.M., Burtcev, A.K. *et al.*, Problems of legal regulation of diagnostics and human genome editing in the russian federation. *Lex Russica*, 6, 71–79, 2019.

62. Stambolsky, D.V., Bryzgalina, E.V., Efimenko, A.Y. *et al.*, Informed consent to the receipt and use of human cellular material: Juristic and ethical regulation. *Russ. J. Cardiol.*, 12, 84–90, 2018.
63. Osborn, L.S., Regulating three-dimensional printing: The converging worlds of bits and atoms. *Campbell Univ. School Law, San Diego Law Rev.*, 51, 553–622, 2014.
64. Bogdanov, D.E., Bioprinting technology as a legal challenge: Determining the model of legal regulation. *Lex Russica*, 6, 80–91, 2019.
65. Degtyarev, Y.G., Fomin, O.Y., Soltanovich, A.V. *et al.*, Informed consent for medical interference: Medical and bioethical and legal issues in 3D bioprinting. *Int. J. Bioprint.*, 6, 3, 2020, 2014. *Zdravookhranenie*, 2, 27–38.
66. Povarov, Y.S., Requirements for consent to conduct research, treatment or diagnostics in the area of human genome. *Jurid J. Samara Univ.*, 5, 23, 2019.
67. European Commission, *European Commission-DG Health and Food Safety and European Medicines Agency Action Plan on ATMPs*, 2017, Available from: https://www.ema.europa.eu/en/documents/other/european-commission-dg health-food-safety-european-medicines-agency-action-planadvanced-therapy_en.pdf.
68. Li, P., 3D Bioprinting: Regulation, innovation, and patents, in: *3D Bioprinting for Reconstructive Surgery*, pp. 217–231, Elsevier, Berlin, 2018.
69. Ayusheeva, I.Z., Problems of legal regulation of contractual relations in the process of creating bio-print human organs. *Lex Russica*, 6, 92–99, 2019.
70. Ksenofontova, D.S., Legal issues of creation and use of bioprinted human organs. *Lex Russica*, 1, 109–118, 2019.
71. Kelly, E., FDA regulation of 3D_printed organs and associated ethical challenges. *Univ. PS Law Rev.*, 166, 515–546, 2018.
72. Vijayavenkataraman, S., Lu, W.F., Fuh, J.Y., 3D bioprinting an ethical, legal and social aspects (ELSA) framework. *Bioprinting*, 1–2, 11–21, 2016.
73. Medrano, J.V., Andrés, M., García, S. *et al.*, Basic and clinical approaches for fertility preservation and restoration in cancer patients. *Trends Biotechnol.*, 36, 199–215, 2018.
74. Vermeulen, N., Haddow, G., Seymour, T. *et al.*, 3D bioprint me: A socioethical view of bioprinting human organs and tissues. *J. Med. Ethics*, 43, 618–624, 2017.

8

# Optimizing 3D Bioprinting Using Advanced Deep Learning Techniques A Comparative Study of CNN, RNN, and GAN

**K. Sujigarasharma[1]*, Sharulatha S.[2], Lawanya Shri M.[1], Gangadevi E.[3] and Rajesh Kumar Dhanaraj[4]**

[1]*School of Information Technology and Engineering, Vellore Institute of Technology, Vellore, India*
[2]*School of Advanced Science, Vellore Institute of Technology, Vellore, India*
[3]*Department of Computer Science, Loyola College, Chennai, India*
[4]*Symbiosis International (Deemed University), Pune, India*

## *Abstract*

3D bioprinting is a rapidly growing field that uses advanced technology to create 3D structures that mimic the complex architecture and function of living tissue. However, optimizing the printing process and the design of structures is challenging. In this chapter, we introduced a combined deep learning model which utilizes the strengths of three different planning: Recurrent Neural Networks (RNN), Convolutional Neural Networks (CNN), and Generative Adversarial Networks (GAN) for 3D bioprinting optimization. We propose the use of publicly available datasets such as the 3D Slicer Medical Image Segmentation dataset, sensor data from the printing process, and the Open Organ Database dataset. The results from this combined model have the potential to boost the value of accuracy, efficiency, and scalability of 3D bioprinting action.

***Keywords:*** 3D bioprinting, convolutional neural networks, recurrent neural networks, generative adversarial networks, 3D bioprinting optimization

**Corresponding author*: sujigarasharma.k@gmail.com

E. Gangadevi, M. Lawanya Shri, Rajesh Kumar Dhanaraj and Balamurugan Balusamy (eds.) Computational Intelligence in Bioprinting, (157–174) © 2024 Scrivener Publishing LLC

## 8.1 Introduction

3D bioprinting is a rapidly growing field that uses advanced technology to create 3D structures that mimic the complex architecture and function of living tissue. Optimization of 3D bioprinting techniques is essential to improve the printing process's accuracy, efficiency, and scalability. Optimizing 3D bioprinting techniques includes integrating multiple technologies such as microfabrication, nanotechnology, and materials science. This integration allows for the creation of more complex and functional structures, such as 3D-printed organs and tissues [1]. Another trend is the use of bio-inks, which are specially formulated materials that mimic the properties of living tissue and support cell growth and differentiation. The general process of 3D bioprinting is shown in Figure 8.1. In addition, the use of artificial intelligence (AI), Machine learning (ML) and particularly Deep Learning, is becoming increasingly popular for optimizing the printing process and design of structures.

There is also a growing trend toward the use of personalized medicine, where 3D bioprinting is used to create customized structures that are tailored to the specific needs of individual patients. Artificial neural networks which utilize a type of Machine Learning (ML) in Deep Learning, that can be used to optimize 3D bioprinting techniques in several ways. One possible application is using deep learning to predict the behaviour of cells and tissues in response to different printing conditions, such as temperature, pressure, and material properties. This can help to optimize the printing

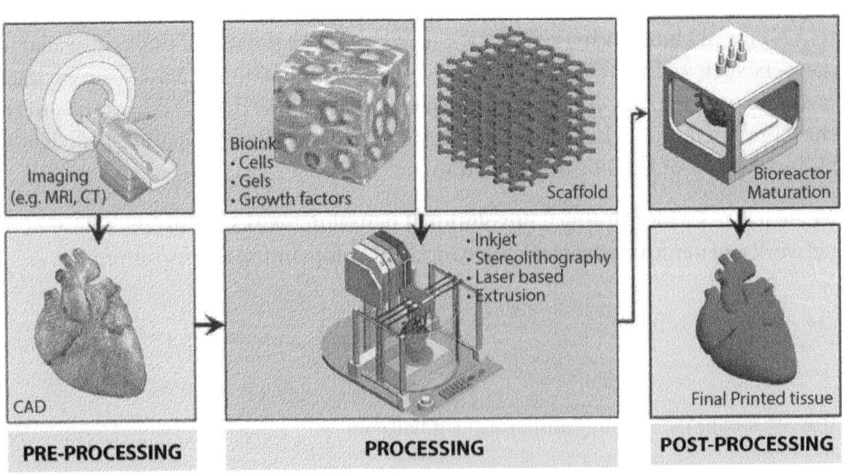


**Figure 8.1** General process of 3D bioprinting.

process and improve the quality of the printed structures. Another application is using deep learning to control and automate the printing process, allowing for greater precision and control. Additionally, deep learning can be used for image analysis and segmentation of 3D printed structures to evaluate their quality and identify any defects.

Deep learning algorithms have been recently used to optimize 3D bioprinting techniques. These algorithms, such as Convolutional Neural Networks (CNN), can be used for image analysis and segmentation of 3D printed structures, to automatically identify and classify different types of cells, tissues, and other features [2]. Recurrent neural networks (RNN) are used to model and predict the behaviour of cells and tissues over time, based on the printing conditions and other factors. Generative adversarial networks (GAN) is can be used to generate new, realistic 3D models of structures, such as orGAN, to be printed. Reinforcement learning: These can be used to control and automate the printing process, allowing for greater precision and control. Other deep learning algorithms such as Autoencoder, Transformer, etc can also be applied to optimize the 3D bioprinting process depending on the specific use case and the dataset. It's worth noting that these are just some examples of deep learning algorithms that can be used to optimize 3D bioprinting this field is still in development and new algorithm and techniques are being proposed regularly.

Generative Adversarial Networks (GAN), Recurrent Neural Networks (RNN), and Convolutional Neural Networks (CNN) can learn from large amounts of data and make predictions or generate new data. Recent trends and advances in 3D bioprinting optimization using deep learning include the use of CNN for image analysis and segmentation, RNN for monitoring and predicting the printing process, and GAN for generating 3D models and design optimization. A range of datasets is being used for deep learning-based 3D bioprinting optimization, including medical imaging datasets for image analysis, sensor data from the printing process, and datasets of 3D models of structures to be printed. The availability of large datasets and advances in computation power has enabled these deep learning-based approaches to be applied to 3D bioprinting. Overall, deep learning-based approaches have the potential to revolutionize 3D bioprinting by enabling the optimization of the printing process and the design of new structures. However, these techniques are still in starting stage of establishment and further to realize their full potential research is needed.

## 8.2 Convolutional Neural Networks in Optimization of 3D Bioprinting

Convolutional Neural Networks (CNN) are a kind of deep learning algorithm which is especially well-matched for picture analysis and segmentation tasks. In the context of 3D bioprinting, CNN can be used to automatically identify and classify different types of cells, tissues, and other features in 3D printed structures. This can be useful for evaluating the quality of the printed structures and identifying any defects. CNN is composed of layers of artificial neurons that are designed to mimic the behaviour of the visual cortex in the human brain. They can process and understand images by applying mathematical operations, such as convolution and pooling, to the input data [3]. The convolution operation is used in the input image extraction, while in the pooling operation, the dimensionality of data is reduced [4]. One of the main advantages of CNN is that they can learn features from the data automatically, without the need for manual feature engineering. This makes them well-suited for tasks such as object detection, semantic segmentation and image classification.

In 3D bioprinting, CNN can be trained on a data set of 3D images of printed structures, such as tissue samples. The network can then be used to classify different types of cells, such as stem cells or mature cells, or to identify defects in the printed structures. CNN can also be used to segment the different parts of the structure, such as the vasculature or the extracellular matrix. It's worth noting that CNN is not the only type of deep learning algorithm that can be used for 3D image analysis and segmentation, other algorithms, such as U-Net and 3D-CNN are also being used in this field and are showing promising results.

## 8.3 RNN in Optimization of 3D Bioprinting

Recurrent neural networks (RNN) are a kind of deep learning algorithm that is especially well-matched for particular tasks involving sequential data, such as time-series analysis [5]. In the context of 3D bioprinting, RNN can be used to model and predict the behaviour of cells and tissues over time, based on the printing conditions and other factors. This can be useful for optimizing the printing process and for predicting the long-term behaviour of the printed structures. Unlike traditional feedforward neural networks, RNNs have a feedback connection, which allows them to maintain a hidden state that can capture information about the previous inputs.

This makes them well-suited for tasks, such as language modelling, speech recognition, and time-series prediction [6].

In 3D bioprinting, RNN can be trained on a dataset of time-series data, such as images of printed structures taken at different time points after printing. The network can then be used to predict the behaviour of cells and tissues over time, based on the printing conditions and other factors. For example, RNN is used to predict how cells will respond to multiple concentrations of growth factors or different mechanical stresses. RNN can also be used to control and automate the printing process. For example, by using the output of the RNN to adjust the printing conditions in real time, it's possible to achieve greater precision and control over the printed structures. It is worth noting that there are various kinds of RNNs, such as gated recurrent unit (GRU) and long short-term memory (LSTM) networks that can be used for different types of sequential data analysis.

## 8.4 Generative Adversarial Networks (GAN) in Optimization of 3D Bioprinting

Generative Adversarial Networks (GAN) are the kind of deep learning algorithm that is particularly well-suited for tasks involving generating new, realistic data, such as images or 3D models. In the context of 3D bioprinting, GAN can be used to generate new 3D models of structures, such as orGAN, to be printed. These generated models can be used for training, testing or even for actual printing [7]. A GAN is composed of two components, a generator and a discriminator. The generator is a neural network which is trained to generate new data, such as images or 3D models, based on a given input. The discriminator is another neural network that is trained to differentiate between real data and generated data. The main two different networks are trained together in a characterized process, where the generator tries to generate data which is indistinct from real data, and the discriminator tries to correctly identify generated data.

In 3D bioprinting, GAN can be used to generate new 3D models of structures, such as orGAN, to be printed. These models can be used for training, testing or even for actual printing. For example, GAN can be used to generate 3D models of specific orGAN, such as the heart or the liver, based on real data. These generated models can be used to train other deep learning algorithms, such as CNN or RNN, for image analysis and segmentation tasks. It's worth noting that GAN are still a relatively new and rapidly evolving field, and more different kind of GAN had been developed for various

**Table 8.1** Comparison table for CNN, RNN, and GAN in the context of 3D bioprinting optimization.

| | CNN | RNN | GAN |
|---|---|---|---|
| **Data type** | 2D/3D images | Time series data | 3D models |
| **Optimization task** | Image analysis and segmentation | Monitoring and predicting the printing process | Generating 3D models of structures to be printed |
| **Advantages** | Good at identifying patterns and features in images | Good at handling time series data and identifying patterns over time | Good at generating realistic, high-quality 3D models |
| **Limitations** | Limited to analysing images and not well-suited for time series data | Limited to time series data and not well-suited for image analysis | Can be difficult to train and may not always generate high-quality models |

kinds of data and tasks [8]. Some examples include Wasserstein GAN, Style GAN, and BigGAN. The comparison of CNN, RNN, and GAN in the context of 3D bio-printing optimization is presented in Table 8.1.

## 8.5 Datasets Used for Optimization of 3D Bioprinting

There are a variety of datasets that could be used for training deep learning models for 3D bioprinting optimization. The specific dataset will depend on the task and the data being analysed.

For example, in the case of image analysis and segmentation tasks, datasets such as the 3D Slicer Medical Image Segmentation dataset and the Medical Segmentation Decathlon dataset could be used to train CNN and other deep learning algorithms. For training GAN to generate 3D models of orGAN, datasets such as the Open Organ Database can be used. For time series data, datasets of images and other data collected over time during the printing process can be used to train RNN. For printed structures' physical and mechanical properties, datasets such as the Mechanical and Physical Properties of 3D Printed Objects dataset can be used.

It's also worth noting that there are several challenges and benchmarks in the field of 3D bioprinting such as the 3D Bioprinting Challenge and The

Bio fabrication and Bioprinting Challenge that release datasets for training deep learning algorithms for 3D bioprinting. It's also possible to use a sequence of various datasets or even to create a custom dataset, depending on the specific task and the available data.

## 8.6 3D Slicer Medical Image Segmentation Dataset

The 3D Slicer Medical Image Segmentation dataset is a dataset used for testing image and training segmentation algorithms. These datasets consist of a variety of medical images, such as CT and MRI scans, along with corresponding manual segmentations of various structures within the images. The dataset is publicly available and can be downloaded from the 3D Slicer website.

The 3D Slicer Medical Image Segmentation dataset contains a variety of different structures, such as organs, bones, and vessels, which are used to train and test algorithms for a broad range of applications, together with organ segmentation, bone segmentation, and vessel segmentation. The dataset is particularly useful for training algorithms for image segmentation in the context of medical imaging, as it contains a wide variety of medical images and corresponding manual segmentations, which can be used to train and test the algorithm.

The 3D Slicer Medical Image Segmentation dataset is widely used in the medical image analysis community and had been used for many publications to estimate the presence of different image segmentation algorithms. The dataset can be used in combination with deep learning algorithms such as U-Net, for image segmentation tasks it is the most popular architecture. The U-Net architecture is particularly well-suited for image segmentation tasks as it can learn to identify and segment objects within an image, even when the objects are not visible.

## 8.7 Sensor Data

Sensor data refers to information collected by sensors, which are devices that can detect and measure physical phenomena such as temperature, light, pressure, humidity, motion, and more. Sensor data is used in a broad range of applications, including in manufacturing, transportation, healthcare, and many other industries.

Sensor data can be collected in different forms, including analogue and digital data. Analog sensor data is typically continuous, such as data collected by a thermometer or a pressure sensor. Digital sensor data is typically discrete, such as data collected by a motion sensor or a binary sensor.

The data collected by sensors is often raw and unprocessed, and it requires further processing and analysis to extract useful information. This can be done using various techniques, including statistical analysis, machine learning, and data visualization. Sensor data can be analysed to detect patterns, trends, and anomalies, and to make predictions about future events.

In the context of 3D bioprinting, sensor data is used to monitor and control the printing process. It can provide information about the temperature, pressure, movement of the printing head, and other parameters that infect the quality of printed design. Sensor data can be used to enhance the precision, efficiency and accuracy of the printing procedure, and to detect and correct errors before they occur.

## 8.8 Open Organ Database Dataset

The Open Organ Database (OOD) dataset is a publicly available dataset that contains 3D models of various human organs. The dataset was created by researchers at the University of Colorado and is intended for use in medical research, particularly in the field of 3D bioprinting. The OOD dataset includes 3D models of organs such as the heart, liver, kidney, and lungs.

The OOD dataset is available in the PLY format, which is a standard file format for 3D models. The dataset includes 3D models of various resolutions, which can be used for different applications. The dataset also includes information such as the organ's volume, surface area, and segmentation of the organ into different regions. The OOD dataset is intended to be used as a benchmark dataset for evaluating the performance of 3D bioprinting algorithms, as well as for the testing and training of ML models. The dataset can be used for tasks such as image analysis and segmentation, quality control, and other tasks that involve analysing 3D images. The dataset is also used concerning the Initial point for creating new 3D models or for improving the resolution of existing 3D models. It's worth noting that The Open Organ Database is a limited dataset and it may not be able to cover all the variations in human anatomy or the specific needs of certain research or medical cases. Additionally, the dataset may not be fully accurate or up-to-date, and it is important to verify any information before using it in any medical application.

## 8.9 Proposed Model

We introduced a combined deep learning model that utilizes the strengths of three different architectures such as RNN Long Short-Term Memory

(LSTM), CNN U-Net, and Wasserstein GAN. The process of the model has shown in Figure 8.2. The CNN U-Net model can be used for image analysis and segmentation of 3D models of structures to be printed. This can be used to identify specific regions of the structure and extract relevant features for further processing. The RNN LSTM model can be used to monitor and predict the printing process. This can be used to detect errors or deviations from the expected printing process and to make adjustments in real-time to optimize the printing process. The Wasserstein GAN model can be used to generate 3D models of structures to be printed and optimise the structure's design. This can be used to generate new designs or to identify potential issues that need to be addressed before printing. The combined model can be used for quality control, taking into account the image analysis and segmentation done by the CNN U-Net, the predictions of the printing process by the RNN LSTM and the design optimization done by

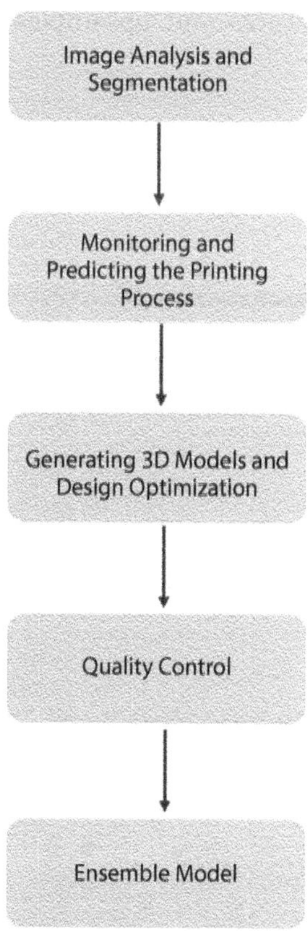


**Figure 8.2** A process of the combined model.

the Wasserstein GAN. The combined model can be used as an ensemble model, where the output of each of the three models (CNN U-Net, RNN LSTM and Wasserstein GAN) are combined and used to make a final decision.

## 8.10 CNN U-Net

The U-Net architecture is a kind of Convolutional Neural Network (CNN) used in image analysis and segmentation tasks. It was originally developed for medical image analysis but it had been applied to various domains [9]. Figure 8.3 U-Net architecture contains two main parts, the decoder and the encoder. The encoder part consists of a sequence of pooling and convolutional layers that reduce the spatial dimensions of the input image are increased while the number of feature maps is increased. This method extracts increasingly higher-level features from the image. The decoder part then takes these features and upsamples them, using transposed

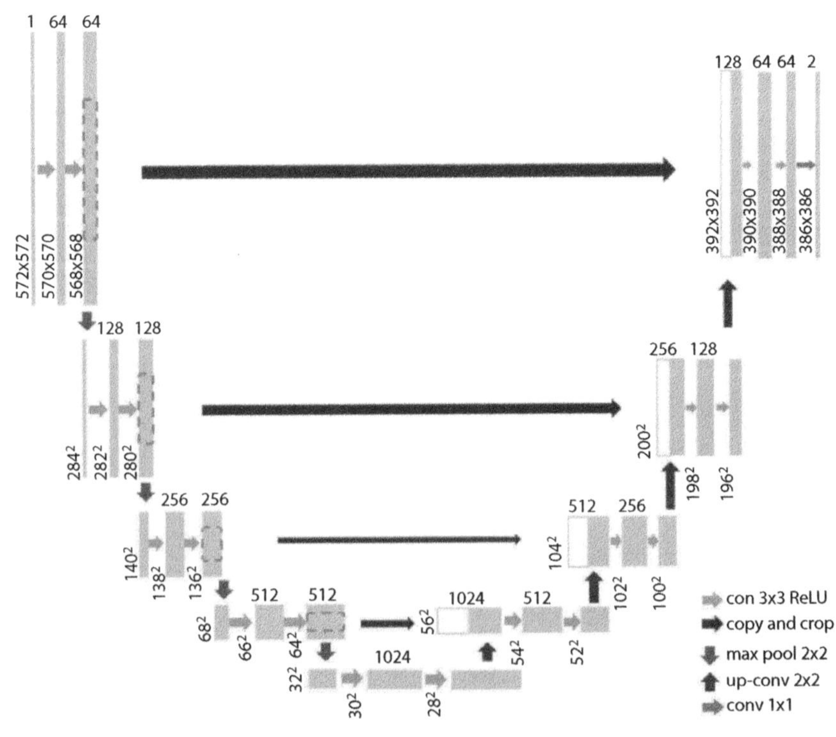


**Figure 8.3** U-Net architecture.

convolutional layers, back to the original image size while increasing the spatial resolution.

The U-Net architecture is called a U-Net because of its shape, which resembles the letter "U". The encoder and decoder are connected by skip connections, which allows data to flow directly from the decoder to the encoder, preserving the spatial information at different scales. This allows the network to make precise predictions for every pixel in the image or picture. The U-Net architecture is trained using an image segmentation task, where the goal is to categorise each pixel in the image into one of several classes [10]. The network is trained by providing it with pairs of figures and their respective segmentation masks. Then the network learns to produce a segmentation mask for a given figure based on its features and the relationships between the pixels in the image. Overall, the U-Net architecture is widely used in medical image analysis and has been successful in tasks such as cell segmentation, organ segmentation, and tissue classification. The tool is effective for picture analysis and segmentation in 3D bioprinting, providing precise predictions and accurate results.

## 8.11 RNN Long Short-Term Memory

Long short-term memory (LSTM) is a Recurrent Neural Network (RNN) architecture well-suited for continuous data analysis, such as the sensor data collected during the 3D bioprinting process. The LSTM architecture has forgotten gates, input, memory cells, and output and that allows it to selectively store and recall information from past time steps [11]. This ability to retain information over a longer time is what makes LSTMs particularly useful for tasks such as predicting the progression of the printing process or detecting anomalies in the sensor data.

In an LSTM network, the memory cells are connected to the input and output layers through the input, output, and forget gates. At each time step, the input and forget gates determine which information should be stored in the memory cells, while the output gate determines which details should be flows to the output layer [12]. In this way, the LSTM can selectively choose which information to retain and which to discard, based on the data it has processed so far.

The LSTM architecture in Figure 8.4 can be fine-tuned for a specific task by adjusting the weights of the connections between the different layers. During training, the LSTM is provided with a pattern of input data and the corresponding output data, and it adjusts its weights to reduce

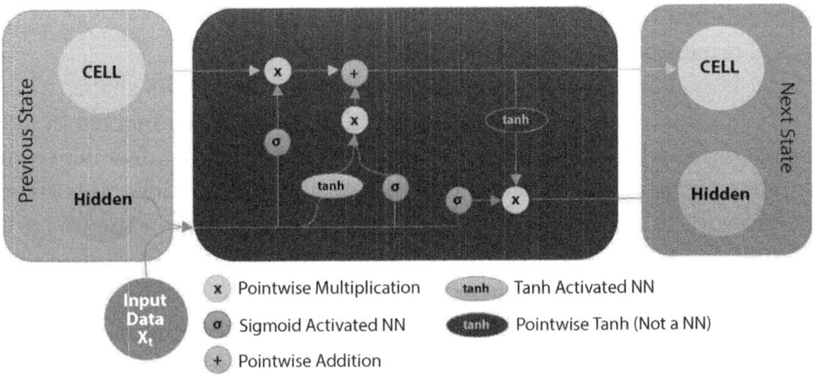


**Figure 8.4** LSTM architecture.

the prediction error. This process is repeated many times until the LSTM reaches a satisfactory level of performance. In the context of 3D bioprinting, LSTMs can be used to monitor and predict the progression of the printing process by analysing the sensor data in real-time. By detecting anomalies or deviations from the expected process, LSTMs can provide valuable information for making decisions about how to adjust the printing process to optimize the outcome.

## 8.12 Wasserstein Generative Adversarial Network

Wasserstein Generative Adversarial Network (WGAN) is a kind of Generative Adversarial Network (GAN) architecture used in deep learning [13]. The main difference between WGAN and other GANs is the use of Wasserstein distance as a measure of the loss function, which helps to overcome the problems of mode collapse and instability during training [14].

The architecture of a WGAN, Figure 8.5 consists of two various neural networks: a generator and a discriminator. The generator uses arbitrary noise as the input and produces a synthetic sample. The discriminator takes both synthetic samples and real ones and outputs a probability score indicating whether the sample is valid or invalid. The objective of the generator is to produce samples that can trick the discriminator into thinking they are real, while the objective of the discriminator is to correctly identify real and fake samples. In the WGAN, the discriminator is trained to reduce the Wasserstein interval between the valid and invalid distributions, which is equivalent to minimizing the Earth Mover's Distance (EMD) between the two distributions. This allows

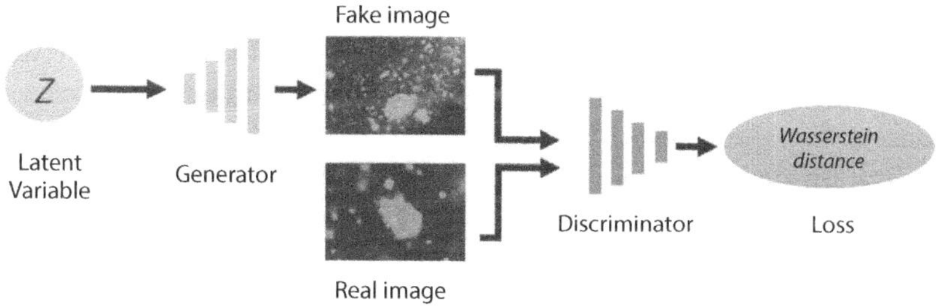


**Figure 8.5** WGAN architecture.

the WGAN to be trained more stably and avoid mode collapse, which is a common problem in traditional GANs. The generator is trained to reduce the Wasserstein intervals between the generated and real distributions. Overall, Wasserstein GANs are an effective tool for inducing high-quality synthetic data, and they have been used in a variety of applications, including 3D bioprinting, computer vision, and natural language processing.

## 8.13 Process of Combined Model

Data Preparation is the first step in optimizing 3D bioprinting using deep learning techniques. We must acquire and prepare the appropriate data to train and test the algorithms. For image analysis and segmentation using a CNN, the 3D Slicer Medical Image Segmentation dataset is publicly available and contains 3D images of various structures such as organs and blood vessels, along with their corresponding segmentation masks. This dataset can be used to train the CNN to identify specific structures in images. For monitoring and predicting the printing process using RNN Long Short-Term Memory (LSTM), sensor data such as temperature, pressure, and movement of the printing head should be collected in real time during the printing process.

Next, we need to choose a network architecture for the optimization of 3D bioprinting. A popular choice for image analysis and segmentation is a U-Net architecture, as it is well-suited for this task and can handle 3D data. The U-Net contains an encoder network, which extracts features from the input image, that generates the segmentation mask. For monitoring and predicting the printing process, a popular choice is the LSTM architecture, as it is well-suited for sequential data and can handle long-term dependencies in the sensor data. The LSTM consists of a sequence of cells that can retain information over time, allowing the network to make predictions based on past observations. Once the data and network architecture have been prepared, the models can be trained using relevant datasets. For the CNN U-Net, the model is trained using the 3D Slicer Medical Image Segmentation dataset, where the model is presented with a figure and its corresponding segmentation mask, and the model's parameters are updated to reduce the distinction between the predicted mask and the ground-truth mask. For the RNN Long Short-Term Memory (LSTM), this model is trained using sensor data, where the model is presented with a sequence of sensor data, and the model's parameters are updated to minimize the discrepancy between the process status as predicted. For pair of two models, the training process helps optimise the predictions' accuracy and precision.

After training the model, it is important to evaluate its performance through testing. This can be done by using a set of images or sensor data that the model has not seen before. The output of the model is then compared to the ground-truth segmentation mask for image analysis or the actual process status for predictive maintenance to measure its accuracy, precision, and recall. This step helps to ensure that the model is functioning correctly and that it can accurately perform the tasks it was designed to do. Once a deep learning model is trained and tested, it can be applied in various ways to optimize 3D bioprinting. The model can be used for quality control by analysing the images of printed structures and detecting defects or errors that may have arisen during the printing process. In addition, the model can also be used for monitoring the printing process in real time, helping to identify potential problems before they occur.

Furthermore, the model can also be applied to predictive maintenance, analysing sensor data over time and predicting when maintenance is required, thereby reducing downtime and prolonging the life of the equipment. Deployment of the combined model of CNN U-Net, RNN long short-term memory (LSTM), and Wasserstein GAN can be accomplished by integrating the model into the 3D printing software or hardware. This allows for real-time analysis of images, sensor data, and quality control. The model can also be used as a standalone tool for quality control, monitoring, and predictive maintenance. After being trained and tested, the model can be effectively utilized in a real-world setting, improving the precision, accuracy, and efficiency of 3D bioprinting. The steps for the combined model of optimizing 3D bioprinting are shown in Figure 8.6.

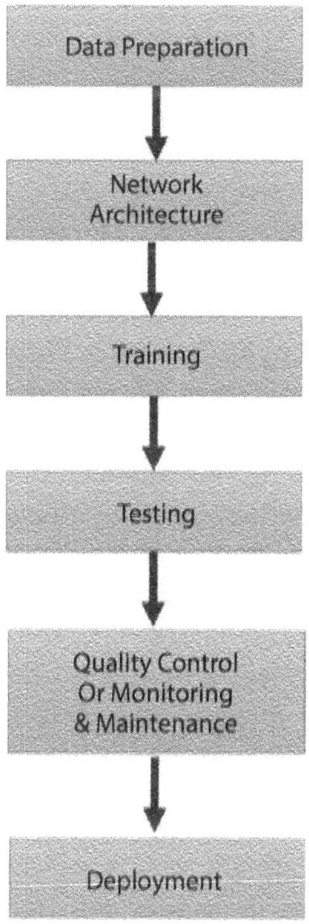


**Figure 8.6** Combined model Steps of optimizing 3D bioprinting.

## 8.14 Conclusion

Advanced deep learning techniques, such as CNN, RNN, and GAN, can be effectively used to optimize 3D bioprinting. CNNs can be used for image analysis and segmentation, quality control, and other tasks that involve analysing 2D images. RNNs, particularly LSTM architecture, are well-suited for monitoring and predicting the printing process, and for predictive maintenance. GANs can be used for generating new 3D models or for improving the resolution of 3D images.

Using these techniques, 3D bioprinting precision, accuracy, and efficiency can all be improved, which can ultimately lead to better outcomes in regenerative medicine, tissue engineering, and other fields. However, it's important to note that the implementation of these techniques requires a good understanding of the 3D bioprinting process, the data, and the deep learning techniques, as well as the ability to evaluate and interpret the results.

## References


1. Zhang, J., Wehrle, E., Rubert, M., Müller, R., 3D Bioprinting of human tissues: Biofabrication, bioinks, and bioreactors. *Int. J. Mol. Sci.*, 22, 8, 3971, 2021. https://doi.org/10.3390/ijms22083971.
2. Jin, Z., Zhang, Z., Shao, X., Gu, G.X., Monitoring anomalies in 3D bioprinting with deep neural networks. *ACS Biomater. Sci. Eng.*, 9, 7, 3945–3952, 2021. https://doi.org/10.1021/acsbiomaterials.0c01761.
3. Brownlee, J., *Deep Learning for Computer Vision: Image Classification, Object Detection, and Face Recognition in Python*, Machine Learning Mastery, Vermont, Victoria, Australia, 2019.
4. Sujigarasharma, K., Rathi, R., Visvanathan, P., Kanchana, R., Emotion-based human-computer interaction, in: *Multidisciplinary Applications of Deep Learning-Based Artificial Emotional Intelligence*, pp. 136–150, IGI Global, Pennsylvania, United States, 2023, https://doi.org/10.4018/978-1-6684-5673-6.ch009.
5. Al-Askar, H., Radi, N., MacDermott, A., Chapter 7-Recurrent neural networks in medical data analysis and classifications, in: *Emerging Topics in Computer Science and Applied Computing, Applied Computing in Medicine and Health*, 1ra. Ed, D. Al-Jumeily, A. Hussain, C. Mallucci, C. Oliver (Eds.), Morgan Kaufmann, Boston, 2016, https://doi.org/10.1016/B978-0-12-803468-2.00007-2.
6. Pascanu, R., Mikolov, T., Bengio, Y.,On the difficulty of training recurrent neural networks, in: *International Conference on Machine Learning, Proceedings of Machine Learning Research*, pp. 1310–1318, May 2013.
7. Ashok, M. and Gupta, A., Comparative study of TRANS-GAN architecture for bio-medical image semantic segmentation, in: *2022 7th International Conference on Communication and Electronics Systems (ICCES)*, IEEE, pp. 1564–1570, June 2022, https://doi.org/10.1109/ICCES54183.2022.9835992.
8. Suganthi, K., Review of medical image synthesis using GAN techniques, in: *ITM Web of Conferences*, vol. 37, EDP Sciences, p. 01005, 2021, https://doi.org/10.1051/itmconf/20213701005.
9. Ronneberger, O., Fischer, P., Brox, T., U-net: Convolutional networks for biomedical image segmentation, in: *Medical Image Computing and Computer-Assisted Intervention–MICCAI 2015: 18th International Conference*, Munich, Germany, October 5–9, 2015, vol. Proceedings, Part III 18, Springer International Publishing, pp. 234–241, 2015.
10. Doyle, S., Forbes, F., Jaillard, A., Heck, O., Detante, O., Dojat, M., Sub-acute and chronic ischemic stroke lesion MRI segmentation, in: *Brainlesion: Glioma, Multiple Sclerosis, Stroke and Traumatic Brain Injuries: Third International Workshop, BrainLes 2017, Held in Conjunction with MICCAI 2017*, Quebec City, QC, Canada, September 14, 2017, vol. Selected Papers 3, Springer International Publishing, pp. 111–122, 2018.

11. Laurent, C., Pereyra, G., Brakel, P., Zhang, Y., Bengio, Y., Batch normalized recurrent neural networks, in: *2016 IEEE International Conference on Acoustics, Speech and Signal Processing (ICASSP)*, IEEE, pp. 2657–2661, March 2016, https://doi.org/10.1109/ICASSP.2016.7472159.
12. Rajeev, R., Samath, J.A., Karthikeyan, N.K., An intelligent recurrent neural network with long short-term memory (LSTM) BASED batch normalization for medical image denoising. *J. Med. Syst.*, 43, 1–10, 2019.
13. Goodfellow, I., Pouget-Abadie, J., Mirza, M., Xu, B., Warde-Farley, D., Ozair, S., Bengio, Y., Generative adversarial networks. *Commun. ACM*, 63, 11, 139–144, 2020. https://doi.org/10.1145/3422622.
14. Jiang, M., Yuan, Z., Yang, X., Zhang, J., Gong, Y., Xia, L., Li, T., Accelerating CS-MRI reconstruction with fine-tuning Wasserstein generative adversarial network. *IEEE Access*, 7, 152347–152357, 2019. https://doi.org/10.1109/ACCESS.2019.2948220.

# 9

# Research Trends in Intelligence-Based Bioprinting for Construction Engineering Applications

**Kuldeep Singh Kaswan[1], Jagjit Singh Dhatterwal[2*], Om Prakash[1], Balamurugan Balusamy[3] and Feslin Anish Mon[4]**

[1]*School of Computing Science and Engineering, Galgotias University, Greater Noida, India*
[2]*Department of Artificial Intelligence & Data Science, Koneru Lakshmaiah Education Foundation, Vaddeswaram, Andhra Pradesh, India*
[3]*Shiv Nadar University, Delhi-National Capital Region (NCR), Greater Noida, India*
[4]*Faculty in Department of Information Technology, University of Technology and Applied Sciences, Ibri, Sultanate of Oman*

## *Abstract*

Bioprinting has emerged as a promising technology in the field of construction engineering, enabling the fabrication of complex and functional biological structures for various applications. In recent years, the integration of intelligence-based approaches, such as artificial intelligence (AI) and machine learning (ML), with bioprinting has gained significant attention, offering new avenues for innovation and advancement. This review aims to provide an overview of the research trends in intelligence-based bioprinting for construction engineering applications. First, we introduce the fundamental principles of bioprinting and its relevance to construction engineering, emphasizing the importance of bioink formulation, printing techniques, and post-printing processes. Subsequently, we delve into the integration of intelligence-based approaches, including AI and ML, into bioprinting workflows. We explore the utilization of AI and ML algorithms for tasks such as material optimization, structural design, process automation, and quality control in bioprinting. Furthermore, this review examines the potential applications of intelligence-based bioprinting in construction engineering. We discuss the fabrication of bio-inspired materials and structures, such as biomimetic scaffolds and living architectural

**Corresponding author*: jagjits247@gmail.com

E. Gangadevi, M. Lawanya Shri, Rajesh Kumar Dhanaraj and Balamurugan Balusamy (eds.) Computational Intelligence in Bioprinting, (175–192) © 2024 Scrivener Publishing LLC

components, using AI-guided design and printing strategies. Additionally, we explore the integration of sensor technologies and real-time feedback systems for intelligent monitoring and adaptive control of the bioprinting process. Moreover, the review highlights the challenges and future prospects of intelligence-based bioprinting in construction engineering. These challenges include the need for extensive data collection, algorithm development, and standardization of processes. The ethical, regulatory, and sustainability considerations related to the use of biologically derived materials in construction are also discussed.

***Keywords:*** Additive manufacturing, 3D bio-printing, computer-aided design, biological tissues

## 9.1 Introduction

By fusing together biological and inorganic material in a preset three-dimensional (3D) spatial layout, bioprinting is a family of AM methods whose purpose is to construct artificial components that imitate the activities of real tissues and organs [1]. Conventional 3D printing or AM is used to produce bioengineered structures with applications in regenerative medicine, tissue engineering, reconstructive surgery, drug discovery, pharmacokinetics, medical, and basic cell-biology research from CAD models representing the fundamental configuration of the target tissue or organ. The printing of living cells within a specific medium called bioink is the most novel aspect of bioprinting, and it presents a number of new challenges, such as how to prevent the degradation of living cells while printing constructs with a 3D volumetric shape similar to that of natural tissues and organs [2].

There is a growing interest in personalized medicine, and as a result, bioprinting technologies have garnered a lot of attention from both academics and corporations in recent years. During the last decade, several innovations in methods and technology related to biomanufacturing have emerged, from specialized 3D bioprinter to one-of-a-kind "raw biomaterials" dubbed bioinks [3].

Bioprinters are a kind of 3D printer that deposit living cells in precise patterns to form complex tissues. Bioprinters often employ bioinks, which are soft biomaterials stuffed with living cells and changed according to specific procedures, to achieve this purpose and produce biological structures. Inkjet printing may benefit from the use of dissolvable secondary materials that can provide vertical support and protection for cells.

Many recent articles have provided overviews of various bioprinting methods and/or bioink classifications. A rigorous and quantitative assessment of the real landscape, which would include the analysis of publications, patents, and corporations, has not been conducted. To our knowledge, there is only one scientometric review dedicated to the topic of 3D bioprinting, and since it was published in 2017, the data it analyzed only goes as far as the middle of 2016. This means that it is out of date and fails to take into account the most recent findings in this field of study. In fact, in 2016, more than two-thirds of all papers on bioprinting were published [4].

By highlighting new trends and the most recent breakthroughs in the area, this research seeks to spark the interest of scientists and professionals already working in more traditional types of 3D printing or AM. This review's rigorous study of the geographic distribution of labs and firms actively participating in 3D bioprinting, as well as their output in terms of publications, patents, innovative tools, and production procedures, may be an important resource for individuals newly joining the area.

The outline for this paper is as follows. In "The academic research trends," we discuss and debate the results of our literature review, including our in-depth analysis of the most prolific authors and the most active research networks throughout the world. You may utilize the "landscape" section to find new and current hubs of technological innovation by reading a description of the market and patent landscape. In the last section, labeled "Conclusions," the most important results are recapped [5].

## 9.2 Analysis of Bioprinting

Analysis of facts on Research Trends in Intelligence-Based Bioprinting for Construction Engineering Applications:

Bioprinting in Construction Engineering: The abstract highlights that bioprinting has gained significant attention in the field of construction engineering. This fact reflects the increasing interest in utilizing bioprinting technology for fabricating complex and functional biological structures relevant to construction applications [7].

Integration of Intelligence-Based Approaches: The abstract emphasizes the integration of intelligence-based approaches, specifically artificial intelligence (AI) and machine learning (ML), with bioprinting workflows. This fact indicates the recognition of the potential benefits that AI and ML algorithms can bring to bioprinting processes in terms of optimization, automation, and quality control.

Material Optimization and Structural Design: The abstract mentions the utilization of AI and ML algorithms for tasks such as material optimization and structural design in bioprinting. This fact highlights the application of intelligent algorithms to enhance the material properties and structural characteristics of bioprinted constructs, enabling the fabrication of bio-inspired materials and structures.

Process Automation and Quality Control: The abstract discusses the role of AI and ML algorithms in automating bioprinting processes and ensuring quality control. This fact indicates the potential of intelligence-based approaches to improve the efficiency, accuracy, and reliability of bioprinting operations, leading to more consistent and reproducible results.

Sensor Technologies and Real-Time Feedback Systems: The abstract mentions the integration of sensor technologies and real-time feedback systems in intelligence-based bioprinting. This fact reflects the incorporation of intelligent monitoring and adaptive control mechanisms into the bioprinting process, enabling real-time adjustments based on feedback data for improved printing outcomes.

Challenges and Considerations: The abstract acknowledges the challenges associated with intelligence-based bioprinting, including the need for extensive data collection, algorithm development, and addressing ethical, regulatory, and sustainability considerations. This fact highlights the importance of addressing these challenges to fully exploit the potential of intelligence-based bioprinting in construction engineering applications.

Overall, the analysis of the facts indicates that intelligence-based bioprinting is an emerging research trend in construction engineering. The integration of AI and ML algorithms offers opportunities for optimizing materials, improving structural design, automating processes, and implementing quality control measures. However, several challenges and considerations need to be addressed for the successful implementation of intelligence-based bioprinting in the construction industry.

Several types of 3D bioprinting technologies are described in various reviews, therefore it is feasible to classify these articles in many ways. Bioprinting, like traditional AM techniques, often involves depositing biomaterial in successive layers. While the bioprinting literature does not necessarily assume the correct nomenclature supplied in the AM standards, AM procedures similar to those used for polymers are routinely employed. Printing resolution and speed in the nozzle-deposition class of technologies vary depending on the precision of the bioprinting head, the size of the nozzle, and the method by which droplets are generated. Optically dependent methods include vat photopolymerization, often known as stereolithography in the bioprinting literature, and its more sophisticated two-photon polymerization cousin [8].

Most studies have focused on extrusion-based bioprinting, perhaps because it is the cheapest method that allows for the use of a wide array of printed biomaterials and the least expensive choice for an entry-level bioprinter. The second and third most common techniques are vat photopolymerization and inkjet bioprinting [9]. The former has several benefits over the latter, including higher resolution, larger cell density, and a broader selection of bioink viscosities. Most bioprinters use a technique called drop-on-demand (DOD) patterning, which has led to the widespread use of jetting for printing finer details [10].

The extrusion-based method has skyrocketed in popularity as a result of the broad availability of low-cost bioprinters in recent years. Vat photopolymerization 3D bioprinting, on the other hand, is swiftly becoming the gold standard for printing complex tissues.

## 9.3 Model Development in Bioprinting Technology

Mini-tissue synthesis for drug screening, illness model development, and regenerative medicine for the ultimate replacement of tissues and organs are some of the most promising applications of 3D bioprinting. Patients' own cells may be harvested via biopsies, the form of the new organ or tissue can be developed, and then the transplant can be performed [11]. As far as we are aware, several hurdles have yet to be overcome, rendering full implementation of this perfect method unattainable. Significant examples include vascularization and multi-material printing. Vital nutrients and oxygen must be transported to living cells through vascularization to sustain their existence.

As in real biological tissues, the 3D framework needs to support the coexistence of several cell types and hydrogels. Progress is being made in three main areas of 3D bioprinting research, all of which are crucial when considering the ultimate goals and driving factors (Figure 9.1):

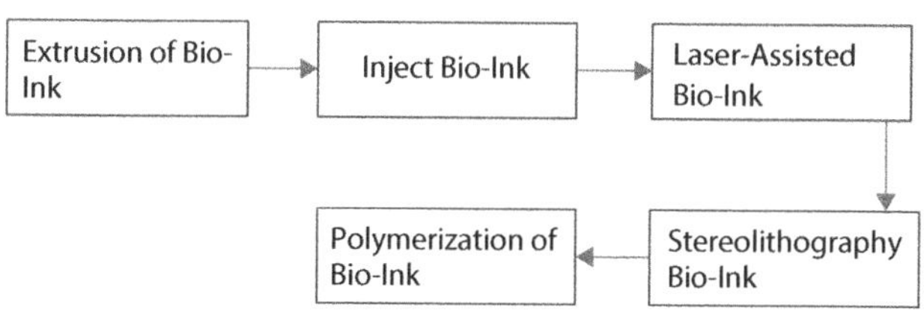


**Figure 9.1** A structure of 3D bioprinting technology.

Study driven by the promise of 3D bioprinting's many potential end uses, such as the fabrication of disease models and organs-on-a-chip for use in drug development. Bioink formulation innovations to improve printability or enable tissue differentiation and maturation, as well as the research of bioprinted cell constructs. In order to improve the resolution and accuracy of 3D bioprinting without harming cells, to make it easier to design complex shapes, to decrease printing times and costs, and to allow for specific functionality, process-driven research of printing technology is required [12]. To highlight the most important advances in the subject, we utilized text analytics keywords to categorize publications published since 2000 into topics. The article count and historical growth for each category are shown.

Several publications, especially review papers, that explain the fundamentals of 3D bioprinting may be classified as process-driven research. Biomimicry, a basic subject, has shown steady growth since 2010, while relatively new ideas like 4D bioprinting have produced 28 scientific publications despite only being available since 2016. While the bioprinting procedure has been described in several published works, the ethical concerns raised by this technology have gotten far less coverage [13].

Over 40% of all articles discussing 3D bioprinting applications include a specific tissue or organ as their first word. The most often studied tissues are typically highlighted in review articles, including bone, cartilage (particularly articular cartilage), vascularized tissue, cardiac tissues, liver, neural tissue, skin, pancreas, cornea, kidney, and muscle. Yet, recent years have seen a rise in curiosity in unexplored topics such as encapsulated T-cells; dental tissue regeneration; nerve repair; the lungs; the intestines; the thyroid; the urethra; and the lungs. There is a good chance that this pattern will persist [14].

Throughout the last decade, graft and implant applications, pathology models, and organs-on-a-chip have all come to play a more significant role (as measured by the ratio of reviews to total articles). Research in this area has traditionally focused on topics like bioglues, grafts, and implants; however, there are now novel approaches, such as the BioPen (a portable device invented by Wallace and co-workers for printing cartilage *in vivo*) and the use of bioprinting in cryopreservation [15].

Using 3D bioprinting, pathology models have been employed for *in vitro* studies of disease from the start. There are several types of cancer that have been characterized using 3D-bio printed models; such examples are breast cancer, mammary ductal carcinoma, appendix cancer, mesothelioma, glioblastoma, and metastases. Bioprinting has been used to create models of a wide range of diseases and conditions, including epilepsy, diabetes, degenerative diseases, immune-enhanced organoids for immunotherapy screening, and wound healing. The pharmaceutical business has been making use of 3D bioprinting for tasks including drug discovery, development, and screening since 2011. One side of this argument is that making sick organs and tissues from patient cells allows for a more personalized approach to medication development. Mini-tissues made in a standardized serial fashion may be useful for high-throughput screening of large pharmacological libraries (drug screening or novel drug discovery) [16]. One day, we want to be able to 3D print patient-specific models using the patient's own cells so that different chemotherapy drugs may be tried *in vitro* to see which one works best. There is little doubt that the implications of translational medicine and 3D bioprinting in regenerative medicine, as well as their clinical translation, will soon become hot topics [17].

Comparing to other applications, publications on translational medicine are relatively young (starting in 2009), with 117 total publications, of which more than 60% are classified as published studies. Nevertheless, the application of 3D-bioprinted tissues in healthcare is still in its infancy, and no organs or tissues printed by 3D bioprinting have been implanted *in vivo* in human patients to the best of our knowledge. In spite of this, 3D printing of biomaterials is gaining popularity in the medical area, notably for the production of bone and dental implants and grafts, and in surgery for the production of patient-specific 3D models on which surgeons may rehearse prior to the actual operation.

Despite the fact that the first 3D printed object appeared in 2004, the most of the critically important research were not published until around 2010. Microfluidics and organs-on-a-chip are two examples where this is particularly relevant. There are presently less than a hundred books dedicated to this topic alone. Although bioprinting may be used to create *in vitro* models for drug screening in the case of pathologic organs, it can also be utilized to study gene expression and cell differentiation under diverse conditions in the case of healthy organs by controlling microfluidics and the microenvironment [18].

One of the most exciting areas of biomaterials research is the development of novel bioinks to achieve certain biological, mechanical, and chemical properties; this accounts for around a quarter of all papers on bioprinting [6]. This field is still developing since bioinks weren't very intriguing until the introduction of 3D bioprinting. Since more information is needed to standardize testing on 3D cultures, the number of reviews on bioprinting has increased. While rheology (21) and structural integrity (22) are receiving more attention, imaging (73 publications), biological characterization (726 papers), resolution (49 publications), and printability

(32 publications) dominate the literature in this domain (9 articles). Recent papers on bioinks have focused on the need of striking a balance between printability and specificity for the intended cell or tissue.

Other from a few fundamental requirements, every kind of cell has extremely distinct requirements (e.g., aqueous environment, sufficient oxygen and nutrient diffusion, appropriate pH, physiological osmolarity of key vitamins and minerals). To multiply and construct their extracellular matrix, several cell types need specialized attachment sites, substrate characteristics, and physical space (ECM). There are both natural and synthetic bioinks on the market, and they each have their own special capabilities when it comes to 3D printing (e.g., bioinks for material extrusion, jetting, or photopolymerization have distinct rheological characteristics, form fidelity, and printability qualities) [19].

Recent publications on 3D bioprinting have focused on several cutting-edge cell types such stem cells, spheroids, and organoids. Differentiation and maturation remain open questions in biology, and researchers are always exploring new cell types in the hopes of finding answers. Skeldon et al. found that the most frequent types of stem cells used in 3D bioprinting were mesenchymal stem cells, neural stem cells, and human induced pluripotent stem cells (iPSCs). We focused on nasal and bone marrow stem cells, but found that human adipose-derived multipotent stem cells (hASCs) had also been employed. For 3D bioprinting, spheroids have been used as the main living component of bioinks since 2003. In the short time since its introduction in 2017, organoids have risen to prominence as a cutting-edge cell source for 3D bioprinting [20].

Characterizing and creating new process approaches for 3D bioprinting get surprisingly little attention in the literature. Just 15% of articles will be written on this topic this year, down from over 30% in 2010. The rise of other scientific disciplines such as biology, medicine, and materials science could explain this trend away from engineering. Second, most techniques used in this field are adapted forms of more common 3D printing techniques that use polymers. Yet, some advanced technologies have been under-researched [21]. The literature reflects this trend, with laser-based techniques (144 papers) and stereolithography (14 reviews) taking center stage (83 articles and 18 reviews). While 3D bioprinting by extrusion began in 2001, it has seen rapid growth after the introduction of commercial bioprinters in 2015. Inkjet is one of the more established techniques, having been introduced in 2006.

Moreover, several of the application categories contain printing techniques that just discover new applications for established printing processes and equipment (e.g., creating channels to form vascularized tissue).

One example of this is suspension bioprinting, which was first developed to create vascularized tissues like skin. Among the most recent methods is freeform reversible embedding of suspended hydrogels (FRESH), which has evolved into FRESH 2.0. In FRESH 2.0, a bioink is extruded in a dissolvable suspension bath typically made of a gelatin microparticle slurry, enabling 3D bioprinting of constructs with higher resolution and being useful for the production of vessels of very small diameters (5 to 10 m). Recently, a whole human heart was 3D printed using this technique. Extrusion bioprinting is used in the technique known as sacrificial writing in functional tissues (SWIFT), which generates a suspension bath of microscopic blood vessels and vascularization within a live, working tissue [22].

Mironov *et al.* 2020 also revealed a novel use for magnets called magnetic levitation (also affiliated to the company 3D Bioprinting Solutions). Initial research shows that the bioinks used in magnetic bioprinters cannot withstand Earth's gravity, despite the technology's promising results. Space agencies like ESA and NASA are investigating the possibility of using microgravity to improve the 3D printing of delicate human tissues like blood vessels and muscles. Hence, an approach that does not use scaffolding or nozzles is required (i.e., without magnetic nanoparticles) [23]. Human space flight and attempts to solve the world's organ scarcity on Earth might be profoundly affected by the advent of bioprinting in space (including Moon and Mars programs). Due to the increased risk of injury on such long journeys, scientists must create a way to 3D print tissues and organs for use by astronauts right away in case of an emergency. One may make the case that 3D bioprinting is a mission enabler for these kinds of endeavors (i.e., space exploration and planet colonization).

## 9.4 3D Bioprinting Academic Institutions in the World

There are several academic institutions around the world that have established themselves as leaders in the field of 3D bioprinting. Here are some prominent institutions known for their contributions to 3D bioprinting research and education:

Wake Forest Institute for Regenerative Medicine (WFIRM) - United States
Harvard University - United States

Massachusetts Institute of Technology (MIT) - United States
University of Oxford - United Kingdom
ETH Zurich - Switzerland
University of Tokyo - Japan
Tsinghua University - China
Peking University - China
Imperial College London - United Kingdom
University of Wollongong - Australia
Queensland University of Technology (QUT) - Australia
Eindhoven University of Technology - Netherlands
Rice University - United States
Stanford University - United States
Johns Hopkins University - United States

These institutions have dedicated research centers, laboratories, and academic programs focused on 3D bioprinting, tissue engineering, and regenerative medicine. They conduct groundbreaking research, collaborate with industry partners, and educate the next generation of scientists and engineers in the field of bioprinting.

It's important to note that this list is not exhaustive, and there are many other institutions globally that contribute to the advancement of 3D bioprinting. The field is rapidly growing, and new institutions are emerging as key players in this area of research and innovation.

While the field of research trends in intelligence-based bioprinting for construction engineering applications is relatively new, there are some academic institutions known for their expertise in 3D bioprinting and construction engineering that may also be involved in this research area. Here are a few academic institutions that specialize in both 3D bioprinting and construction engineering, which could potentially explore the intersection of these fields:

Singapore Centre for 3D Printing (SC3DP) at Nanyang Technological University, Singapore
Institute of Biotechnology and Bioengineering (IBB) at the University of Lisbon, Portugal
Biofabrication and Tissue Morphology Group at University College London (UCL), United Kingdom
Advanced Manufacturing Technology Research Centre at the University of Sheffield, United Kingdom
Tissue Engineering and Bioprinting Laboratory at RMIT University, Australia

Department of Architecture, Design, and Media Technology at Aalborg University, Denmark
Advanced Tissue Biofabrication Centre (ATBC) at Queensland University of Technology (QUT), Australia
College of Architecture and Design at New Jersey Institute of Technology (NJIT), United States
Department of Civil, Structural and Environmental Engineering at University at Buffalo, United States
School of Architecture at Tsinghua University, China

While these institutions have expertise in 3D bioprinting and construction engineering, it is important to note that the specific research focus on intelligence-based bioprinting for construction engineering applications may vary. Researchers within these institutions may be exploring the intersection of these fields or have the potential to expand their research into this area.

## 9.5 Emerging Bioprinting Technology

Bio-printing, also known as 3D bioprinting, is an emerging technology that has gained significant attention in recent years. It has the potential to revolutionize various industries, including healthcare, pharmaceuticals, and tissue engineering. From a market perspective, bio-printing presents both opportunities and challenges.

### 9.5.1 Opportunities

Tissue Engineering and Regenerative Medicine: Bio-printing enables the fabrication of complex, functional tissues and organs, which holds great promise for tissue engineering and regenerative medicine. It offers the potential to create patient-specific implants, organ-on-a-chip models for drug testing, and tissue constructs for transplantation.

Pharmaceutical Research and Drug Testing: Bio-printed tissue models can provide more accurate and physiologically relevant platforms for drug discovery and testing. This technology allows for the creation of realistic tissue models that mimic human physiology, enabling more efficient and accurate drug development processes.

Personalized Medicine: Bio-printing has the potential to advance personalized medicine by enabling the production of patient-specific tissues and organs. This can help in creating customized implants, prosthetics, and even organs for transplantation, reducing the risk of rejection and improving patient outcomes.

Veterinary Medicine: The application of bio-printing is not limited to human medicine. It can also benefit veterinary medicine by enabling the creation of animal-specific tissues and organs, which can improve veterinary surgeries and treatments.

### 9.5.2 Challenges

Complexity and Cost: Bio-printing involves multiple components, including bioinks, printers, and post-processing equipment. The technology is still in its early stages, and the cost of equipment and materials can be high, making it challenging to scale up and make it accessible for widespread use.

Regulatory Hurdles: The regulatory environment surrounding bio-printing is complex and evolving. The safety and efficacy of bio-printed tissues and organs need to be thoroughly evaluated and regulated to ensure patient safety. The regulatory process can be time-consuming and costly, potentially slowing down the market adoption.

Ethical Considerations: The ability to print human-like tissues and potentially entire organs raises ethical questions and considerations. The use of human cells and tissues for bio-printing raises concerns about consent, intellectual property, and equitable access to healthcare.

Standardization and Quality Control: Ensuring consistent quality and reproducibility of bio-printed tissues and organs is a challenge. Standardizing the printing processes, bioink formulations, and post-processing techniques is crucial to ensure reliable and predictable outcomes.

## 9.6 Development in Bioengineering

The increasing global presence of companies producing commercially viable 3D bioprinters is indicative of the rapid development of both AM and the bioengineering industry. The number of businesses that provide bioprinting and bioink services is growing as the technology develops. In reality, major bioprinting companies are trying to break into the market by using novel technology approaches.

It is the goal of most companies to create all-inclusive extrusion-based bioprinting platforms with features such as the ability to print with a wide range of materials (such as viscous pastes, gels, hydrogels, ABS/PLA and other filaments or polymer powders, liquids, ceramics, and foods), and a variety of tools (such as a laser system for ultra-high-precision cutting and engraving, a CNC milling machine, a photo-crosslink LED.

Companies that invest time and resources into developing unique answers also populate this scene. One possible starting point is the use of tissue spheroids and methods for manipulating them, such as magnetic bioprinting using tools like the Organ. A magnetic bioprinter called Aut was transported to the ISS by the Russian company 3D Bioprinting Solutions on the Soyuz MS-11. To further facilitate the development of scaffold-free tissues, Japanese company Cyfuse Biomedical developed a platform to regulate spheroids using the Kenzan bioprinting technique. The formation of 3D structures is made possible by placing cellular spheroids in a temporary array of needles using a robotic cell-dispensing device. Other companies, such as Germany's Cellbricks, however, choose to develop complex 3D-printed cell culture systems utilizing a proprietary, non-public variant of stereolithography.

Several companies have even suggested higher-degree bioprinters as a way to increase system flexibility and print feature breadth. The US company Advanced Solutions is one such example; they have invented a bioprinter arm that uses robotic extrusion and can load as many as 10 different biomaterials at once. Several companies have chosen to focus on less common uses for their 3D bioprinters, such as the scaffold-free printing of 3D cylinders using a proprietary bio-ink to construct containers by the Chinese company Revotek. The company claims they transplanted a small segment of abdominal artery into 30 rhesus monkeys with good results.

The bioprinting sector is being propelled by more than just extrusion-based devices. One of the many technologies currently under development with the goal of achieving the precision of single-cell deposition is Image Based Single Cell Isolation (IBSCI), developed by the French company Cellenion. IBSCI is a high-resolution-based technology that uses automated image acquisition, processing, and advanced algorithms to automatically isolate single cells from a cell suspension. Another French business, Poietis, uses their own PIATM software in tandem with laser-assisted bioprinting, extrusion-based technology, and inkjet technology to recreate a 3D model of a full tissue. To print bioinks with a coaxial shell, microfluidic platforms with a onprinthead crosslinking device are being developed by companies like Aspect Biosystems in Canada.

Newer companies are forming partnerships with more established ones to increase product variety and decrease pricing. A firm like CELLINK provides a wide range of choices, including inexpensive bioprinters (both extrusion- and DLP-based) and a number of specialized bioinks. The Holograph XTM is one of the first systems to use two-photons stereolithography, and it was created in partnership with Prellis Biologics. Bioprinting speeds of up to 250,000 voxels per second are possible because to the utilization of multiphoton technology, which employs a parallel set of photons to treat millions of locations in the bio-ink at simultaneously.

Leading companies like Organovo are diversifying their revenue streams by offering services and products in addition to their proprietary bioprinting technology (such histologically and functionally identical liver and kidney tissue models).

In addition, the 2015 biotech company BIOLIFE4D out of Illinois is one to watch (USA). The company's long-term objective is to use a patient's own cells to 3D-print a heart that is both a perfect match and completely functioning for the patient. In 2018, BIOLIFE4D created a vascularized and contractile iPSC cardiac patch that could be 3D-bioprinted. They claim that their next significant achievement in 2019 will be the manufacture of a human mini-heart, a 3D-bioprinted replica of a tiny heart.

## 9.7 Evolution of Patent Trends in Bioprinting

The evolution of patent trends in bio-printing provides insights into the technological advancements, innovation landscape, and intellectual property protection within the field. While I don't have access to real-time patent data, I can provide a general overview of how patent trends in bio-printing have evolved over the years based on historical information up to September 2021.

Early Patents (2000s): In the early 2000s, bio-printing technology was still in its infancy, and patent activity in the field was relatively low. Patents during this period mainly focused on foundational technologies and methods related to bioprinting, including printer designs, bioink formulations, and scaffold materials.

Expansion of Patent Activity (2010s): The 2010s witnessed a significant increase in patent filings related to bio-printing. This growth can be attributed to advancements in bioprinting techniques, materials, and applications. Patents during this period covered a wide range of areas, including tissue engineering, regenerative medicine, drug screening, and organ-on-a-chip technologies.

Multidisciplinary Approach: As bio-printing evolved, patents began to reflect the multidisciplinary nature of the field. Innovations spanned various disciplines, including biomedical engineering, materials science, cell biology, and computer-aided design. Patent filings often involved combinations of technologies, such as bio-printing with microfluidics, nanotechnology, or stem cell research.

Materials and Bioinks: Patents related to bioink formulations and materials used in bio-printing have seen significant growth. Researchers and companies have focused on developing bioinks that mimic natural tissues, provide structural support, and enable cell viability and functionality. Patents in this area cover novel biomaterials, gelatinous constructs, hydrogels, and composite materials.

Specific Applications: Patents have also emerged in specific bio-printing applications, such as skin tissue engineering, bone regeneration, vascular constructs, and cartilage repair. These patents often focus on specific methods, designs, or compositions tailored to address the unique requirements of a particular tissue or organ.

Automation and Robotics: With advancements in automation and robotics, patents have emerged in the area of intelligent systems and robotic-assisted bio-printing. These patents often cover technologies that enable precise positioning and layer-by-layer deposition of bioink, as well as real-time monitoring and feedback control.

Intellectual Property Landscape: The evolution of patent trends in bio-printing has led to an expanding intellectual property landscape. Companies, research institutions, and universities have been actively filing patents to protect their innovations and secure competitive advantages in the field. This has also led to collaboration and licensing agreements to access patented technologies.

## 9.8 Conclusions

In conclusion, research trends in intelligence-based bioprinting for construction engineering applications are evolving rapidly, driven by the potential of integrating artificial intelligence (AI) and machine learning (ML) with bioprinting processes. The integration of intelligence-based approaches in bioprinting workflows offers new avenues for innovation and advancement in the field. The research trends discussed in this context include material optimization, structural design, process automation, and quality control. AI and ML algorithms are being utilized to enhance these aspects of bioprinting, leading to improved material properties, optimized structural designs, automated processes, and reliable quality control mechanisms. Furthermore, the potential applications of intelligence-based bioprinting in construction engineering are diverse. It includes the fabrication of bio-inspired materials and structures, such as biomimetic scaffolds and living architectural components, using AI-guided design and printing strategies. Additionally, the integration of sensor technologies and real-time feedback systems enables intelligent monitoring and adaptive control of the bioprinting process. However, several challenges and considerations need to be addressed to fully exploit the potential of intelligence-based bioprinting in construction engineering applications. These challenges include the need for extensive data collection, algorithm development, standardization of processes, as well as addressing ethical, regulatory, and sustainability concerns. Overall, intelligence-based bioprinting holds great promise for revolutionizing construction engineering by enabling the fabrication of advanced bio-inspired materials and structures. Continued research and development efforts are crucial to overcome the challenges and further advance the capabilities of intelligence-based bioprinting in construction engineering applications.

## References


1. Lee, J.M., Sing, S.L., Yeong, W.Y., Wiria, F.E., Bioprinting in cardiovascular tissue engineering: A review. *Int. J. Bioprint.*, 5, 2, 197, 2019.
2. Baliyan, A., Dhatterwal, J.S., Kaswan, K.S., Jain, V., Role of AI and IoT techniques in autonomous transport vehicles, in: *AI Enabled IoT for Electrification and Connected Transportation*, pp. 1–23, Springer Nature Singapore, Singapore, 2022.
3. Ozbolat, I.T. and Hospodiuk, M., Current advances and future perspectives in extrusion-based bioprinting. *Biomaterials*, 76, 321–343, 2016.
4. Atala, A. and Murphy, S.V., 3D bioprinting of tissues and organs. *Nat. Biotechnol.*, 32, 8, 773–785, 2014.
5. Dhatterwal, J.S. and Kaswan, K.S., Securing big data using big data mining, in: *Data Driven Decision Making Using Analytics*, pp. 1–14, CRC Press, Boca Raton, 2021.
6. Groll, J., Burdick, J.A., Cho, D.W., Derby, B., Gelinsky, M., Heilshorn, S.C., Mironov, V., A definition of bioinks and their distinction from biomaterial inks. *Biofabrication*, 11, 1, 013001, 2018.
7. Berwal, P., Dhatterwal, J.S., Kaswan, K.S., Shashi, K., *Computer applications in engineering and management*. CRC Press, 2022.
8. Bhattacharjee, T., Zehnder, S.M., Rowe, K.G., Jain, S., Nixon, R.M., Sawyer, W.G., Angelini, T.E., Writing in the granular gel medium. *Sci. Adv.*, 4, 6, eaar2767, 2018.
9. Mandrycky, C., Wang, Z., Kim, K., Kim, D.H., 3D bioprinting for engineering complex tissues. *Biotechnol. Adv.*, 34, 4, 422–434, 2016.

10. Munoz-Abraham, A.S., Rodriguez-Davalos, M.I., Bertacco, A. *et al.*, 3D printing of organs for transplantation: Where are we and where are we heading? *Curr. Transpl. Rep.*, 3, 1, 93–99, 2016. https://doi.org/10.1007/s40472-016-0089-6.
11. Dhatterwal, J.S., The role of MAS based CBRS using DM techniques for the supplier selection. *Int. J. Comput. Sci. Eng.*, 7, 5, 1658–1665, 2019.
12. Jakus, A.E., Rutz, A.L., Shah, R.N., Advancing the field of 3D biomaterial printing. *Biomed. Mater.*, 11, 1, 014102, 2016.
13. Derakhshanfar, S., Mbeleck, R., Xu, K., Zhang, X., Zhong, W., Xing, M., Asghari, F., 3D bioprinting for biomedical devices and tissue engineering: A review of recent trends and advances. *Bioact. Mater.*, 3, 2, 144–156, 2018.
14. Moroni, L., Boland, T., Burdick, J.A. *et al.*, Biofabrication: A guide to technology and terminology. *Trends Biotechnol.*, 36, 4, 384–402, 2018.
15. Gao, G., Kim, B.S., Jang, J. *et al.*, Recent strategies in extrusion-based three-dimensional cell printing toward organ biofabrication. *ACS Biomater. Sci. Eng.*, 5, 3, 1150–1169, 2019. https://doi.org/10.1021/acsbiomaterials.8b00691.
16. Kaswan, K.S., Dhatterwal, J.S., Kumar, S., Lal, S., Cybersecurity law-based insurance market, in: *Big Data: A Game Changer for Insurance Industry*, pp. 303–321, Emerald Publishing Limited, Bingley, England, 2022.
17. Groll, J., Boland, T., Blunk, T. *et al.*, Biofabrication: reappraising the definition of an evolving field. *Biofabrication*, 8, 1, 013001, 2016.
18. Salaris, F., Rosa, A., Stabile, F.M., 3D printing techniques for biofabrication of tissue engineering scaffolds. *J. Healthc. Eng.*, 2017, 9410438, 2017.
19. Mironov, V., Boland, T., Trusk, T. *et al.*, Organ printing: computer-aided jet-based 3D tissue engineering. *Trends Biotechnol.*, 21, 4, 157–161, 2003.
20. Gbureck, U., Hölzel, T., Doillon, C.J., Printing tissue-mimicking phantoms with a multi-material 3D bioprinting system. *J. Ceram. Sci. Technol.*, 9, 4, 515–520, 2018.
21. Jakab, K., Norotte, C., Marga, F. *et al.*, Tissue engineering by self-assembly and bio-printing of living cells. *Biofabrication*, 2, 2, 022001, 2010.
22. Keriquel, V., Guillemot, F., Guillotin, B., High-throughput bioprinting of form-degrading scaffolds for tissue engineering. *Biofabrication*, 2, 1, 014104, 2010.
23. Rivenson, Y., Zhang, Y., Günaydın, H. *et al.*, PhaseStain: the digital staining of label-free quantitative phase microscopy images using deep learning. *Light Sci. Appl.*, 9, 1, 1–10, 2020.

# 10

# Design and Development to Collect and Analyze Data Using Bioprinting Software for Biotechnology Industry

**Jagjit Singh Dhatterwal[1*], Kuldeep Singh Kaswan[2], Sanjay Kumar[3], Balamurugan Balusamy[4] and Lakshmana Kumar Ramasamy[5]**

[1]*Department of Artificial Intelligence & Data Science, Koneru Lakshmaiah Education Foundation, Vaddeswaram, Andhra Pradesh, India*
[2]*School of Computing Science and Engineering, Galgotias University, Greater Noida, India*
[3]*Department of Computer Science & Engineering, Galgotias College of Engineering & Technology, Greater Noida, India*
[4]*Shiv Nadar University, Delhi-National Capital Region (NCR), Delhi, India*
[5]*Faculty of Computer Information Systems, HCT Higher Colleges of Technology, United Arab Emirates*

***Abstract***

The rapid advancements in biotechnology have led to an increased need for sophisticated data collection and analysis systems to support research and development processes. In recent years, bioprinting technology has emerged as a promising tool in the field of tissue engineering and regenerative medicine. This abstract focuses on the design and development of a data collection and analysis system using bioprinting software for the biotechnology industry. The proposed system aims to streamline the process of collecting and analyzing data generated during bioprinting experiments, enabling researchers to gain valuable insights and accelerate scientific discoveries. The system integrates bioprinting hardware, software, and data analysis techniques to create an efficient and comprehensive solution. To design the data collection aspect, the system incorporates sensors and imaging devices to capture critical parameters during the bioprinting process. These sensors may include temperature sensors, pressure sensors, and cameras to monitor the printing environment and ensure precise control. Additionally, the system can collect data on material properties, such as viscosity and elasticity, which impact the bioprinted

**Corresponding author*: jagjits247@gmail.com

E. Gangadevi, M. Lawanya Shri, Rajesh Kumar Dhanaraj and Balamurugan Balusamy (eds.) Computational Intelligence in Bioprinting, (193–210) © 2024 Scrivener Publishing LLC

structures' quality and functionality. For data analysis, advanced algorithms and statistical models are employed to extract meaningful information from the collected data. Machine learning techniques can be applied to predict the optimal bioprinting conditions, evaluate the performance of different materials, and identify trends or patterns within large datasets. Furthermore, the system facilitates the visualization of data through interactive interfaces, aiding researchers in interpreting and presenting their findings effectively. The benefits of this data collection and analysis system are manifold. Firstly, it enables real-time monitoring of the bioprinting process, allowing researchers to intervene promptly and adjust printing parameters as needed. Secondly, the system provides a comprehensive dataset for subsequent analysis, fostering a deeper understanding of the bioprinting process and its outcomes. Finally, the integration of bioprinting software with advanced data analysis techniques offers the potential for optimization and improvement of bioprinted structures, leading to advancements in tissue engineering and regenerative medicine.

***Keyword:*** Bioprinting, tissue improvement, cells interaction, modifying crosslinking

## 10.1 Introduction

Through "layer by layer precisely positioning of biological materials, biochemical and live cells, with spatial control of the placement of functional components," bioprinting is a computer-assisted method capable of assembling tissues. The purpose of bioprinting is to create artificial tissues and organs in a controlled, efficient, and robotic setting [1]. As pointed out in a recent piece, bioprinting is only one of two basic methodologies within biofabrication. When "live single cells, bioactive compounds, biomaterials, or cell-aggregates tiny enough to be printed" are used in the production of a construct, the result is known as bioprinting. The necessity for a standardized advanced manufacturing technique in bioprinting, as suggested by these criteria, became clear. Establishing engineering methods that can imitate biological input via design and material choices is essential. As cells and materials undergo remodeling after manufacturing, it is crucial to recreate the passage of time during bioprinting.

Integrating technology and fields like developmental biology, stem cells, computers, and materials science into computer-aided tissue engineering and regenerative medicine is essential. Comprehensive engineering strategies for the development of engineered tissue may be established via the integration of many disciplines. Yet, the development of design concepts and tactics for effective 3D bioprinting is complicated by the interplay of different technologies and disciplines. While developing for various tissue engineering applications, it is important to establish a relationship between factors, concerns, and printing methods for the bioprinting process [2].

The present state of bioprinting technology and its possible applications have been detailed in several academic articles. These domains include stem cell studies, cancer research, heart valve replacement, skin replacement, etc. Bioprinting staples like methods, materials, and cell banks have been covered as well. Function dictates the specific materials and cells used in the construction of synthetic tissues. In its place, this research provides a holistic evaluation of 3D bioprinting's process capabilities and design approaches [3].

We have identified and investigated a wide range of bioprinting design factors, including shape and resolution, material heterogeneity, and the dynamic interaction between cells and their surroundings. The features, capacities, and traits of a process are shaped by these design decisions. Bioprinting considerations such as data processing, bioink composition, and process compatibility are discussed. Some of the bioprinting techniques that have been researched and described include direct bioprinting, in-process crosslinking, post-process crosslinking, indirect bioprinting, and hybrid bioprinting. Tissue engineering has the potential to make use of a wide variety of bioprinting techniques. The potential of organ bioprinting is explored, along with its future directions. Future bioprinting design approaches may benefit from this in-depth analysis [4].

## 10.2 Digital Technology in Bioprinting

With the use of algorithms, engineered tissues may be manufactured with more precision in terms of cell-materials shape formation, multi-material depositing, and the inclusion of varied cell populations. Since bioprinting is a digital technique, it may be combined with computational analysis to learn about and simulate the dynamic processes involved in the development of engineered tissues before, during, and after their manufacture [5]. When detailing the possibilities of bioprinting and addressing different design elements throughout the material formulation and technology selection phase, a design map is necessary. In particular, evaluating bioprinting requires various design considerations [6].

### 10.2.1 Shape of Bioprinting

By carefully regulating the bioprinting process, precise anatomical geometries may be created. Printing designs that have some kind of anatomical likeness as a freestanding construct has recently been shown. For instance, in an effort to bioprint a heart valve using an extrusion-based technology, Duan *et al.* presented a reverse engineering strategy. First, the micro-CT was used to take pictures of a human aortic valve that had been removed from a pig [7]. The input data was utilized by an open-source additive manufacturing process to make an anatomically perfect heart aneurysm. In a second illustration, Duan and colleagues demonstrated an alternative design method in which heart valves are sketched in Solidwork and then imported for printing. Being a technique that can print any digital structure, bioprinting is useful for creating a 3D construct that is anatomically similar to the desired tissue or organ.

In 3D bioprinting, the resolution and precision depend heavily on the smallest material unit created. Similarly, with additive manufacturing, the printing step is not where resolution and accuracy are optimized [8].

**Table 10.1** Bio printing medical accuracy.

| | **Var subject** | **Explanation** |
|---|---|---|
| Form and conclusiveness | Files (.STL, .DATA) Resolution, precision, and size of the printing platform | Parallels in structure Forms of matter, such as globules, filaments, and patches. Amount of layers |
| Heterogeneity of materials | • Diverse cell kinds • Single cells • Cell aggregates (also known as tissue spheroids) | • Autologous, Allogeneic, or Xenogeneic Cell Type Stem Cells or Specialized Cells as Cell Sources |
| | • Gelatin, gelatin, gelatin, alginate, fibrin, and their derivatives are all examples of hydrogel biomaterials.<br>• * Thermoplastics such PLGA, PCL, PLA, etc. | • Adsorbate concentration Print durability Viscoelasticity Technological integrity after manufacturing Processes window Hardening method and velocity |
| | Growth factor | • Formation of patterns and pinpoint location |
| The active nature of genetic components remodeling | Biological Processes<br>• Fusion and sorting of cells<br>Alterations in material, deterioration, and form (shrinkage or expansion)<br>Interaction of Cells and Materials [16]<br>• Reorganization of cells | Transitions in Structure<br>Time's Changing |

In Table 10.1, the layer thickness of a bioprinted structure also affects its accuracy. Other printing variables include the print path's elevation, distance between print nodes, and nozzles size. might impact the layer thickness as well. If you print several layers in the z-axis, the final product may look quite different from the original blueprint [6].

## 10.2.2 Heterogeneity Units of Material

### *10.2.2.1 Tissue Improvement*

Both in the field and in the lab, co-culture systems have various applications in biology. A wider variety of cells in the population may lead to better results in tissue engineering via processes including cell proliferation, DNA synthesis, and extracellular matrix (ECM) creation. Coordinated deposition of two or more cellular components in a biomaterials project is limited by the qualities of the starting materials [7]. With bioprinting, scientists can create almost any kind of material part they can imagine. Every droplet-based printing technique—from inkjet printing to electrodynamic jetting to acoustic droplet ejection to laser-induced forwards transfer (LIFT)—uses the same basic unit. Droplets or filaments of substance are often extruded when using an extrusion-based printing method. Stereolithography and digital light processing are two examples of techniques that use photopolymer cured in vat polymerization, and their resolution depends on the size of the laser spot or beam area. Details of problems handled utilizing these techniques are provided [8]. When it comes to biochemical and physical function, the resolution generated tissue goes much beyond the diameters of droplets and strands. Transplantation of these co-cultured tissues into animal models yielded improved biological outcomes for the host tissue. Human umbilical vein endothelial cells (HUVECs) and human mesenchymal stem cells were patterned onto a polyester urethane urea (PEUU) patch by Gaebel *et al.* using laser-induced forward transfer (LIFT) to create a cardiac patch matrix [9]. The number of blood vessels increased after the printed patch was implanted into the myocardium of a rat with myocardial infarction. Human umbilical vein endothelial cells (HUVECs), human neonatal dermal fibroblasts (HNDFs), and 10T1/2 murine fibroblasts were used to print intricate heterogeneous structures. Nonetheless, the intricate musculoskeletal system's connective tissue interface has been documented. A single construct composed of regions with different cell types and mechanical qualities [10] recaptures the complex cell-matrix interaction between the cell-rich muscle side and the ECM-rich tendon side.

#### *10.2.2.2 Formation of Biomaterials*

The modular nature of bioprinting machines means that many materials may be dispensed from a single printer. Sequential deposits of fibronectin and thrombin and sodium alginate and calcium chloride produce fibrin and calcium alginate, respectively, two kinds of quickly cross-linking hydrogel. Hydrogels may be generated faster by bioprinting due to the decreased distance between the hydrogel precursor and its crosslinker [11].

Manipulating the site of biomaterial deposition might provide light on the complex mediated activation. Since bioprinting permits the precise synthesis and variable manufacturing of biomaterials, it was used to make a high throughput array for investigating cellular-biomaterial interactions. Certain forms of biomaterial were shown to attract cells more than others [12]. To investigate how heterogeneity tissues may be made, we first printed a single construct in which the amount and type of decellularized extracellular matrix (dECM) varied in different regions. Using what we learn from studying how cells interact with biomaterials, we can tailor the biological response to a given printing technique by careful biomaterial selection and formulation [13].

Dispensing several components in a deliberate method, for instance, may increase the dimensional integrity of a bio printed object (foreword). Researchers have taken use of bioprinting's ability to use several materials to print a tubular vascular system with a build/support geometry. The combination of biodegradable hydrogels with thermoplastic prior to tissue formation also generated a sufficient mechanical scaffold [14]. When printing a muscle-tendon unit, for instance, Merceron *et al.* use thermoplastic polyurethane for the musculature and poly(epsilon-caprolactone) for the tendons since the former is more elastic (PCL). A tendon is generally rigid and short-lived at its bone insertion, while a muscle's elasticity is mechanically exploited throughout contraction and relaxation. Two polymers and two cell types were printed next to one another to provide the interfacial area necessary to fabricate the complicated muscle-tendon unit [15].

#### *10.2.2.3 Biomaterial and Biological Factors*

The chemical and physical gradients present in the native cellular microenvironment in embryonic biology are responsible for inducing a cascade of cell behaviors. In order to stimulate a specific cellular response, bioprinting has been utilized to design biomaterials and biological components. Cellular migration inside a hydrogel may be facilitated by bioprinting via the controlled deposition of growth agents. Graduating multilayers of growth factors at two separate areas allowed for the development of a multiphase complicated tissue structure at the tendon-bone contact [16].

### 10.2.3 Dynamic Changes in Fabrication Process

Tissue maturation is a necessary step after bioprinting and may take either *in vitro* or *in vivo*. Even after the manufacturing process is complete, bioprinted tissue continues to alter dynamically. These alterations are detectable on a cellular and material level [17]. During tissue maturation, cellular processes including fusing and sorting alter the form of bioprinted tissue. Self-assembly and spheroid fusion are two hallmarks of scaffold-free bioprinting, a technique used to create 3D tissues from scratch. The printed spheres mature into a cylindrical structure with a lumen like a blood vessel. In addition to the aforementioned morphological alterations, printed cells also repair the original hydrogel matrix by secreting natural ECM, such as collagen and glycosaminoglycan. Comparable dynamic changes, such as swelling and deterioration, occur in the hydrogel used in bioprinting, which might compromise the bioprinted product [18].

To aid in the development of artificial tissue, a novel bioreactor has been created. To condition the cultured construct biochemically or mechanically, A bioreactor is an engineered system that allows for the precise manipulation of physical measurements and public transportation rates. Bioprinted tissue spheroids need for a unique kind of bioreactor, one that uses a combination of dripping irrigation and triple perfusion to grow the tissue in a controlled environment [19].

## 10.3 Designing Techniques in Bioprinting

Data preparation, bioprinting technology selection, and bioink formulation are the first steps in the bioprinting process (cells and biomaterials). These material and technological decisions are reflective of pre-printing, during-printing, and post-printing design concerns in bioprinting [20].

### 10.3.1 Data Processing in Biomedical Imaging

For their bioprinting projects, researchers have used two distinct methods. The first method relies on a toolpath derived from data gleaned through biomedical imaging (reverse engineering). The toolpath for a bioprinter is designed with help from medical imaging data like CT scans [21]. The alternative method involves creating anatomical models in a CAD program like SolidWorks and then sending those files to a bioprinter. Because these models aren't based on anything already in existence, the designers have more leeway in terms of how they shape the product [22].

The engineered structure's geometrical shape will be determined by the blueprint used to print it. Data processing methods used in additive manufacturing are also common in bioprinting setups. In order to adjust the settings of a variety of bioprinters, STL or BITMAP pictures of computer-aided designs are translated into numerical control (NC) code [23].

### 10.3.2 Process Bioprinting Techniques

Bioink must have certain properties and printability for each bioprinting technique. The first need is a bioprinting technique that can successfully print cells. Orifice-based bioprinting techniques, such as extrusion-based and inkjet-based techniques, need consideration of size disparities between cell types. Cell spheroids have recently emerged as a promising tool for tissue engineering [24]. Single cells in suspensions are on the order of 10-30 microns in size, whereas the diameter of a cell aggregation may range from 150 microns to 500 microns. The size of the orifice (the aperture through which the bioprinting material is driven) is an important factor in choosing between extrusion, inkjet, and laser-assisted bioprinting [25]. When utilizing a technique that calls for a hydrogel with certain mechanical qualities, such as printing, that hydrogel should be selected. Certain methods, like jetting, can only be used with certain types of suspension, including those with a low viscosity. The material's surface tension and viscosity are crucial factors in jetting processes. The viscosity and gelation process of hydrogel place constraints on extrusion-based techniques that involve very viscous materials [26].

### 10.3.3 Interaction of Bioink Formulation

The capacity to print anything will depend on the biomaterials you choose and how you formulate them, in this case hydrogel. An in-depth analysis published only recently has shed insight on the molecular ideas underpinning printable hydrogel. Hydrogels for 3D printing need an understanding of basic principles such as colloidal science, bridging processes, and electrostatic attraction. In the next section, we'll discuss why it's crucial to understand the molecular chemistry and connecting process of hydrogel before designing a printing strategy [27].

**Figure 10.1** Interaction of bioink formulation.

The bioprint's intended usage will determine the kind of cell employed. Figure 10.1 shows how stem cells and proliferation and differentiation of a specific lineage have been used in printing technology. One alternative method for enhancing cell adherence is gene transfection. Nevertheless, cells may be transplanted either as a single answer or as a cluster of cells (spheroids) [28].

## 10.4 3D Bioprinting

Several methods of printing and crosslinking have been employed to improve print quality. One or more of these methods were likely used during bioprinting to provide a robust bioprinted structure. There must be a harmony between the rheology of the biomaterial and the design of the printing toolpath for the deposition of biomaterials and cells if a 3D bioprinted construct is to retain its structural integrity throughout printing [29].

### 10.4.1 Optimized Bioprinting

With direct material printing, the desired form and arrangement are printed in advance. A material's or bioformulation ink's is fine-tuned to get the required shear recovery speed before, during, and after publishing is dependent on a material's viscous, yield stress, and biomechanical properties. Making a semi-crosslinked slurry and adding a thinning agent are two ways to improve direct printing's efficiency. Many methods exist for partially crosslinking the hydrogel in order to raise the bioviscosity of the ink. These methods include the inclusion of an enzymatic crosslinker, altering the temperatures of the functionalized hydrogel, and pre-exposing the polymer solution to UV radiation. Hydrogels with improved viscoelastic materials and a storage modulus $G'$ that is equivalent to its loss modulus $G''$ were produced when Skardal *et al.* exposed the polymer solution to UV light prior to imprinting. Extrusion bioprinting requires a bioink with a certain viscosity, and thickening agents including hydroxyapatite particles, nanocellulose, xanthum gum, and gellan gum may assist accomplish [30].

### 10.4.2 Modifying Crosslinking

Due to the possibility of crosslinking occurring during manufacture, hydrogels using a quick gelation process, such as alginate and fibrin, are often employed. During manufacture, the hydrogel antecedent and copolymerization are coaxially extruded by either dumping the predecessor into a cross - linking agent solution or modifying the extruded head. An alternative method involves planning the toolpath of the printer to deposit the precursor hydrogel and crosslinker in a certain order [31].

### 10.4.3 Multiple Crosslinking

Such a printing method often employs a combination of hydrogels with different crosslinking mechanisms. The combination includes Figure 10.2 illustrates the use of two materials, one that improves printability and form consistency during printing, and another that ensures structural fidelity after printing through crosslinking. For instance, the printability was drastically enhanced by combining gelatin-methacrylamide (GelMA) and hyaluronic acid (HA) as opposed to using each component alone. GelMA's cross-linking aids in maintaining the structure post-printing, while HA boosts the stickiness of the combination [32].

### 10.4.4 Enhance Bioprinting

In order to bioprint a structure with a high degree of form accuracy, a build/support setup is often utilized. The build components relate to the customized tissue object of study (i.e., cells and/or hydrogel), while the support materials provide mechanical strength to preserve the structure.

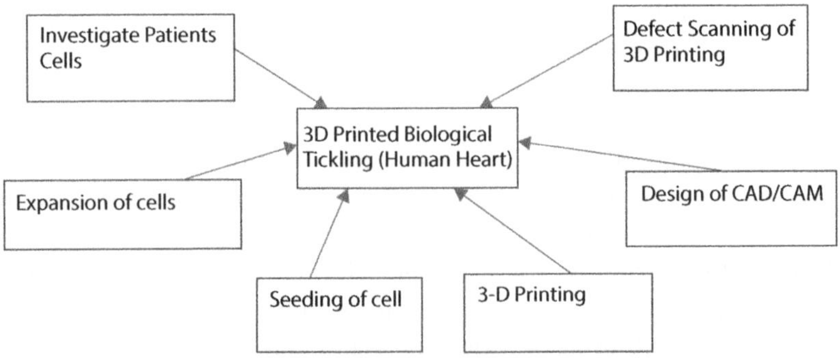


**Figure 10.2** Structure of bioprinting.

Whether it was deposited next to the construction material or in a separated bath, insulation material is often removed in image acquisition. Hydrogel crosslinker added to the support bath before the sacrificial bath is removed improves print quality. Commonly utilized hydrogels for this purpose include gelatin, Pluronic F127, and agarose because to their reversible crosslinking process.

### 10.4.5 Hybrid Bioprinting

Now that bioprinting can be computer-aided, it may be coupled with other manufacturing processes. For instance, multi-scale components may be made by combining bioprinting techniques (for example, extrusion and inkjet) with melt-plotting and electro spinning machinery. Thermoplastic poly (lactic-co-glycolic acid) (PCL) is often used as a substratum for hydrogel in hybrid printing (PLGA). Cell inclusion into the printing process is not conceivable due to the high higher temperature necessary for these thermoplastic materials. Cells are implanted into the elastomer using hydrogel as a transport.

## 10.5 Enhanced Biotissue Printing

By bioprinting, we hope to solve the issue of an inadequate supply of transplantable organs. With the use of computers, it may soon be possible to create artificial organs and tissues. Bioprinted tissues and organs may have their vasculature and growth factors strategically placed on them. This helps improve host-engineered tissue integration. The mechanical properties of newly developed materials may be manipulated by careful formulation and patterning. At the early stages of bioprinting research, computational models will be a useful tool.

### 10.5.1 Integrating Thickness of Engineered Tissue

When moving from the laboratory to the living organism, host-engineered tissue integration is crucial. Diffusion limit sets the upper bound on the maximum thickness of synthetic tissues at around 200 micrometers. The tissue-engineered architecture must have interconnected porosity for proper nutrition and waste exchange. Using bioprinting, microchannels may be easily planned and printed right into the structure. For instance, Kolesky *et al.* used to deposit HNDFs, HUVECs, and 10T1/2 murine fibroblast in an extrusion-based bioprinter using a four-syringe cartridge arrangement to make maximum operation microchannels mimicking a vascular network. Facilitating cell infiltration into a gap between native and artificial tissue is another aspect of integration. Tumor necrosis factor patterning, as shown by Miller and colleagues, may be used to infer that homogenous distribution of exogenous growth elements may be as successful as complex ion concentration in triggering cell migration. Native cells will infiltrate the manufactured tissue in response to transcription factors that have been appropriately implanted.

### 10.5.2 Integration and Enhancement of Cellular Interaction

There has been a lot of research on improving hydrogel viscosity and gelation mechanism to aid bioprinting manufacturing. Modifying the characteristics of a biomaterial may lead to more than simply an enhanced printing experience; it can also trigger a targeted biological reaction. For instance, hydrogel's architecture and atmosphere may be tweaked to affect cell destiny. As human pluripotent embryonic stem cells may undergo self-renewal, transformation, and tissue creation, researchers can influence all of these processes by tailoring the hydrogel's architecture. By removing alginate ingredients from the alginate-collagen framework, stem cells may be segregated in the direction of a specific lineage, and the hydrogel framework can be adjusted to enhance stem cell renewal capacities.

Hydrogel architectural control also includes the formation of hydrogel with graded porosity for promoting cell migration. The hydrogel's graded porosity may be attributed to the diffusion of cellulose enzyme throughout the material. In situations of uniform chemoattractant dispersion, cells migrate toward larger holes because of structural variations in the hydrogel (i.e., no chemoattractant gradient).

### 10.5.3 DNA with a Smart Biomaterial

#### *10.5.3.1 Biomaterials*

Using self-assembly, scientists have found a technique to micro-organize biomaterials. The goal is to create a biopolymer from simpler building blocks like amino acids and DNA. Hydrogel assembly may be guided in a specific direction using an adhesive made from a synthetic DNA biopolymer. The alteration of the cell surface has also been employed as a guide during the construction of cells using DNA templates. In order to facilitate Combining the location control of cell-based therapies with the guidance of DNA glue may be achievable via the coordinating and orientation-directed assembly of building elements consisting of cells and nanomaterials.

#### *10.5.3.2 Reactive Hydrogel to External Stimuli*

Using bioprinting technology and stimulus-responsive hydrogel, it is possible to create synthetic tissue that can change its shape. Some of the most prevalent kinds of interacting hydrogels include mechano-, chemo-, pH-, heat-, and light-responsive. Hydrogels with varying reactions to stimuli may be carefully placed to generate bioprinted constructions with shape-shifting abilities. In addition, hydrogel breakdown may be controlled by integrating cell-responsive regions. As an example, cell growth leads to the degradation of matrix metalloproteinase (MMP)-modified chitosan. Instead, electrically sensitive hydrogel has been developed by boosting the conductivity of alginate with nanowires and carbon nanotubes.

### 10.5.4 Simulation

Scaffold fabrication in AM is being boosted by computer modeling. Mechanical and porosity gradients are possible in scaffold design. In a similar vein, computational modeling will serve as a means to enhance the design of artificial tissue. Simulation software, when coupled with bioprinting, will be a powerful tool for creating synthetic tissue with desirable characteristics. Using computer-simulated simulations, for instance, to optimize the structure of the vascular network may improve the efficacy with which nutrition and wastes are carried. The intra-organ vascular tree may be designed using either computer-generated models or a circuit design method to retrieve data from clinical imaging.

Computer program may be utilized to foresee systemic changes and product characteristics of a bioprinted structure in addition to the creation of designs outlined above. Understanding the synergistic reaction of cells and materials may help provide more accurate simulation results. By simulating the bioprinted construct's performance in a lab, we can estimate when it will reach full functionality upon implantation.

## 10.6 Conclusion

The design and development of a data collection and analysis system using bioprinting software for the biotechnology industry offers numerous benefits and holds significant potential for advancing scientific research and development in tissue engineering and regenerative medicine. The integration of bioprinting hardware, software, and data analysis techniques creates an efficient and comprehensive solution that streamlines the bioprinting process and enables researchers to gain valuable insights from the collected data. By incorporating sensors and imaging devices, the system enables real-time monitoring of critical parameters during the bioprinting process, ensuring precise control and allowing researchers to intervene promptly when necessary. The collection of comprehensive datasets facilitates in-depth analysis, enabling the extraction of meaningful information and the identification of trends and patterns within the data. The application of advanced algorithms and statistical models further enhances the analysis, providing valuable insights and aiding in the optimization and improvement of bioprinted structures.

## 10.7 Future Work

While the design and development of a data collection and analysis system using bioprinting software represents a significant step forward, there are several avenues for future work and improvement:

Enhanced Sensor Integration: Further development of the system should focus on expanding the range of sensors used for data collection. This may include the incorporation of additional parameters, such as pH levels, oxygenation, and cellular behavior, to provide a more comprehensive understanding of the bioprinting process.

Integration with Bioreactors: Bioprinting often involves the use of bioreactors to provide a controlled environment for cell growth and tissue formation. Future work should explore the integration of the data collection and analysis system with bioreactors, enabling simultaneous monitoring of the printing process and the cellular response within the bioreactor.

Data Sharing and Collaboration: The ability to share and collaborate on bioprinting data can significantly accelerate research progress. Future work should focus on developing platforms or frameworks that allow researchers to securely share and analyze data, fostering collaboration and knowledge exchange within the biotechnology industry.

Automation and Artificial Intelligence: The integration of automation and artificial intelligence (AI) techniques can further enhance the efficiency and accuracy of data collection and analysis. Future work should explore the use of AI algorithms to automate data processing, identify anomalies or outliers, and predict optimal bioprinting conditions based on historical data.

Scaling and Industrial Implementation: As bioprinting technology advances, the scalability and industrial implementation of the data collection and analysis system become crucial. Future work should address the challenges of scaling up the system to handle larger volumes of data and adapt it for industrial settings, where high-throughput bioprinting is performed.

## References


1. Zhai, R. *et al.*, 3D bioprinting: A promising method for the fabrication of tissue engineered cartilage and bone constructs. *Front. Bioeng. Biotechnol.*, 8, 573–590, 2020. [Online]. Available: https://www.frontiersin.org/articles/10.3389/fbioe.2020.00397/full.
2. Daly, A.C. *et al.*, 3D bioprinting of tissues and organs. *Nat. Biotechnol.*, 37, 7, 773–785, 2019. [Online]. Available: https://www.nature.com/articles/s41587-019-0114-7.
3. Kaswan, K.S., Dhatterwal, J.S., Kumar, A. (Eds.), *Swarm Intelligence: An Approach from Natural to Artificial*, John Wiley & Sons, Nature Portfolio (UK), 2023.
4. Malda, J. *et al.*, 25th anniversary article: Engineering Hydrogels for biofabrication. *Adv. Mater.*, 25, 36, 5011–5028, 2013. [Online]. Available: https://onlinelibrary.wiley.com/doi/10.1002/adma.201302042.
5. Liu, Y. *et al.*, Design considerations for bioprinting in tissue engineering. *ACS Biomater. Sci. Eng.*, 3, 7, 1120–1137, 2017. [Online]. Available: https://pubs.acs.org/doi/10.1021/acsbiomaterials.7b00133.
6. Dhatterwal, J.S., Kaswan, K.S., Kumar, N., Telemedicine-based development of M-health informatics using AI, in: *Deep Learning for Healthcare Decision Making*, p. 159, 2023.
7. Kaswan, K.S., Gautam, R., Dhatterwal, J.S., Introduction to DSS system for smart cities, in: *Decision Support Systems for Smart City Applications*, pp. 53–76, 2022.
8. Derakhshanfar, R. *et al.*, 3D bioprinting for biomedical devices and tissue engineering: A review of recent trends and advances. *Biofabrication*, 8, 2, 144–156, 2016. [Online]. Available: https://iopscience.iop.org/article/10.1088/1758-5090/8/2/022001.
9. Kang, S.H. *et al.*, A 3D bioprinting system to produce human-scale tissue constructs with structural integrity. *Nat. Biotechnol.*, 34, 3, 312–319, 2016. [Online]. Available: https://www.nature.com/articles/nbt.3413.
10. van Delden, M.C.M. *et al.*, Design and development of advanced 3D-printed bioreactors for tissue engineering. *Tissue Eng. Part B Rev.*, 26, 2, 151–167, 2020. [Online]. Available: https://www.liebertpub.com/doi/10.1089/ten.TEB.2019.0298.

11. Mohamad, A.O. *et al.*, 3D bioprinting technologies for tissue engineering: A review. *J. Saudi Chem. Soc.*, 23, 5, 533–545, 2019. [Online]. Available: https://www.sciencedirect.com/science/article/pii/S131961031830637X.
12. Cheah, C., Chua, C., Leong, K., Cheah, C., Interconnected pore structure of 3D porous tissue engineering scaffolds. *Mater. Sci. Eng. C*, 23, 5, 611–620, 2003.
13. Dhatterwal, J.S. and Baliyan, A., Digital flow-based cyber-physical microfluidic biochips, in: *Cyber Physical Systems*, pp. 1–14, Chapman and Hall/CRC, Broken Sound Parkway, NW, 2023.
14. Patra, P., Rijal, G., Talukdar, S., Additive manufacturing in biofabrication: Current outlook and future directions. *Int. J. Bioprint.*, 4, 2, 143–157, 2018.
15. Yeo, M., Lee, J., Chun, W., 3D bioprinting of organs-on-a-chip. *Bioengineering*, 7, 4, 131, 2020.
16. Kolesky, D., Homan, K., Skylar-Scott, M., Lewis, J., Three-dimensional bioprinting of thick vascularized tissues. *Proc. Natl. Acad. Sci.*, 113, 12, 3179–3184, 2016.
17. Ozbolat, I. and Hospodiuk, M., Current advances and future perspectives in extrusion-based bioprinting. *Biomaterials*, 76, 321–343, 2016.
18. Skardal, A. and Atala, A., Biomaterials for integration with 3-D bioprinting. *Ann. Biomed. Eng.*, 43, 3, 730–746, 2015.
19. Guillotin, B., Souquet, A., Catros, S., Duocastella, M., Pippenger, B., Bellance, S., Laser assisted bioprinting of engineered tissue with high cell density and microscale organization. *Biomaterials*, 31, 28, 7250–7256, 2010.
20. Wang, Z., Abdulla, R., Parker, B., Samanipour, R., Ghosh, S., Kim, K., A simple and high-resolution stereolithography-based 3D bioprinting system using visible light crosslinkable bioinks. *Biofabrication*, 7, 4, 045009, 2015.
21. Dhatterwal, J.S., Baliyan, A., Prakash, O., Reliability driven and dynamic re-synthesis of error recovery in cyber-physical biochips, in: *Cyber Physical Systems*, pp. 15–34, Chapman and Hall/CRC, Broken Sound Parkway, NW, 2023.
22. Tseng, H., Gage, J., Shen, W., 3D biofabrication using stereolithography for biology and medicine. *Engineering*, 1, 2, 261–268, 2015.
23. Koch, L., Deiwick, A., Schlie, S., Michael, S., Gruene, M., Coger, V., Skin tissue generation by laser cell printing. *Biotechnol. Bioeng.*, 109, 7, 1855–1863, 2012.
24. Xu, T., Jin, J., Gregory, C., Hickman, J., Boland, T., Inkjet printing of viable mammalian cells. *Biomaterials*, 26, 1, 93–99, 2005.
25. Chen, Y.C., Lin, R.Z., Qi, H., Yang, Y., Bae, H., Melero-Martin, J.M., Khademhosseini, A., Functional human vascular network generated in photocrosslinkable gelatin methacrylate hydrogels. *Adv. Funct. Mater.*, 22, 10, 2027–2039, 2012.
26. Patra, P., Dey, T., Mishra, S., Das, S., Ghosh, S., Kundu, S.C., Bioink for 3D bioprinting: Recent advances and future prospects. *Biomater. Sci.*, 8, 20, 5424–545, 2020.

27. Fedorovich, N.E., De Wijn, J.R., Verbout, A.J., Alblas, J., Dhert, W.J., Three-dimensional fiber deposition of cell-laden, viable, patterned constructs for bone tissue printing. *Tissue Eng. Part A*, 14, 1, 127–133, 2008.
28. Kolesky, D.B., Truby, R.L., Gladman, A.S., Busbee, T.A., Homan, K.A., Lewis, J.A., 3D bioprinting of vascularized, heterogeneous cell-laden tissue constructs. *Adv. Mater.*, 26, 19, 3124–3130, 2014.
29. Ouyang, L., Highley, C.B., Rodell, C.B., Sun, W., Burdick, J.A., 3D printing of shear-thinning hyaluronic acid hydrogels with secondary cross-linking. *ACS Biomater. Sci. Eng.*, 2, 10, 1743–1751, 2016.
30. Derakhshanfar, S., Mbeleck, R., Xu, K., Zhang, X., Zhong, W., Xing, M., 3D bioprinting for biomedical devices and tissue engineering: A review of recent trends and advances. *Bioact. Mater.*, 3, 2, 144–156, 2018.
31. Ng, W.L., Lee, J.M., Zhou, M., Chen, Y., Tan, Y.N., Tan, N.S., Metal-based antimicrobial strategies for biomedical applications. *Adv. Sci.*, 5, 9, 1700890, 2018.
32. Mazzocchi, A.R., Rajan, S.A., Votanopoulos, K., II, Hall, A.R., Skardal, A., Inman, W., Multiscale frameworks for advancing metabolic models of cancer. *IEEE Trans. Biomed. Eng.*, 66, 11, 3115–3125, 2019.

11

# Cyborg Intelligence for Bioprinting in Computational Design and Analysis of Medical Application

**Kuldeep Singh Kaswan[1]*, Jagjit Singh Dhatterwal[2], Naresh Kumar[3], Balamurugan Balusamy[4] and Gangadevi E.[5]**

[1]*School of Computing Science and Engineering, Galgotias University, Greater Noida, India*
[2]*Department of Artificial Intelligence & Data Science, Koneru Lakshmaiah Education Foundation, Vaddeswaram, Andhra Pradesh, India*
[3]*Applied Computational Science and Engineering, GL Bajaj Institute of Technology & Management, Greater Noida, India*
[4]*Computer Science and Engineering, Shiv Nadar University, Delhi, India*
[5]*Department of Computer Science, Loyola College, Chennai, India*

## *Abstract*
The convergence of cyborg intelligence and bioprinting technologies has paved the way for innovative advancements in the computational design and analysis of medical applications. This abstract explores the integration of cyborg intelligence, which combines the capabilities of artificial intelligence (AI) and human intelligence, with bioprinting techniques to revolutionize the field of medical applications. Cyborg intelligence harnesses the power of AI algorithms and human expertise to enable intelligent decision-making in the design and analysis of medical applications. By utilizing AI algorithms, vast amounts of medical data, including patient-specific information, can be processed to generate optimized designs for various medical devices, implants, and prosthetics. These designs can then be translated into physical objects using bioprinting techniques. Bioprinting, an emerging technology that enables the fabrication of three-dimensional biological structures, plays a crucial role in the realization of these optimized designs. It allows for the precise deposition of bioink materials layer by layer, creating complex and functional tissues, organs, or scaffolds. The integration of cyborg

**Corresponding author*: Kaswankuldeep@gmail.com

E. Gangadevi, M. Lawanya Shri, Rajesh Kumar Dhanaraj and Balamurugan Balusamy (eds.) Computational Intelligence in Bioprinting, (211–238) © 2024 Scrivener Publishing LLC

intelligence enhances the bioprinting process by providing intelligent insights and analysis, optimizing the design parameters and ensuring the fabrication of personalized medical solutions. Furthermore, the computational design and analysis facilitated by cyborg intelligence and bioprinting enable virtual testing and validation of medical applications. By utilizing simulations and computational models, researchers and clinicians can assess the performance, functionality, and compatibility of the printed structures before physical implementation. This accelerates the development process, reduces costs, and enhances patient safety. The integration of cyborg intelligence and bioprinting in medical applications has significant implications across various fields, including tissue engineering, regenerative medicine, and prosthetics. Tissue-engineered constructs can be precisely designed and bioprinted to mimic the native tissues, enabling the regeneration of damaged or diseased organs. Customized prosthetics and implants can be fabricated to match the specific anatomical requirements of patients, leading to improved comfort and functionality. However, challenges such as ethical considerations, data privacy, and regulatory frameworks must be carefully addressed to ensure responsible and secure implementation of cyborg intelligence and bioprinting in the medical field.

***Keywords:*** 4D bioprinting technology, shape memory polymers, tissue engineering regenerative machine, computer-aided design, computed tomography

## 11.1 Introduction

3D printing, a kind of additive manufacturing (AM), has found a number of medical applications, notably in the fields of regenerative medicine, organ and tissue fabrication, and drug delivery. The technology gathers information about specific people by means of imaging modalities including computed tomography (CT) and magnetic resonance imaging (MRI). With this data, CAD models are developed, and eventually, 3D-printed products are made [1]. Many are intrigued by the possibility of using 3D printing to create customized biomedical equipment for particular patients. While biomedical devices created with 3D printing technology have a high degree of physiologic similarity, they are essentially static and not designed to endure dynamic situations. The natural tissue environment of a living organism helps generate tissues and controls biological mechanisms. In addition to previously defined requirements, bioactive materials would benefit from achieving the rigorous standards of exceptional biodegradability as well as biocompatibility [2]. Scientists have developed 4D printing technology to solve this problem. Four-dimensional printing, in which advanced materials morph over a period of virtual time, was first suggested in 2013 by Professors Tibbits and Jerry. 4D printing is a state-of-the-art manufacturing technique that represents the technological next step after 3D printing. The excellent quality, precision, accuracy, and performance enhancements of 4D printing make it possible to create any complex item from a variety of materials, unlike 3D printing techniques. To construct 3D things, 4D printing use 3D printing technology to deposit polymer nanocomposites or other elements layer by layer. Other names for this technique include shape-morphing materials and active origami systems. Bioprinters and the chemicals they employ should be carefully considered [3]. An external stimulus, such as water, heat, light, a magnetic field, an electric field, a small organic molecules, a certain pH, the traction force of a cell, or an enzyme, may modify the structure of a sophisticated applications at any point in the 4D screen printing. There are a variety of 4D printing techniques now in use, including stereolithography (SLA), fused deposition modeling (FDM), inkjet printing (IJP), digital light processing (DLP), selective laser sintering (SLS), micro-extrusion, and direct ink writing (DIW). It provides a synopsis of the different 4D printing techniques used in the production of these substances [4].

The advent of cutting-edge technologies like 4D printing has facilitated breakthroughs in several disciplines, including aviation, prototyping, biotechnology, and medicine. Twenty people lose their lives per day while waiting for a transplantation, according to studies conducted all around the world. 4D printing is revolutionizing the healthcare business and improving people's lives in many ways. Just a few examples are tissue engineering (TE), scaffolds, and dentistry. The transplantation of any organ or tissue formation becomes crucial in the case of an accident. This is why researchers are exploring the possibility of creating artificial organs and tissues using actual human cells. The use of bioprinters to create exact replicas of stem cells and organs marks the beginning of the bioprinting technology [5].

Some of the variable micro-environment tissue engineering (TE) disciplines where 4D printing technology has shown substantial advancement so far include cardiac tissue, capillary density, muscle tissue, neural tissue, bone tissue engineering (BTE), and the fabrication of vascular stents. Using biocompatible substances that promote the body's own healing mechanisms, tissue engineers, and regenerative medicine (TERM) practitioners want to maintain or restore to tissue repair. Scaffolds provide mechanical integrity, human cells aid in the recovery period, and biochemical, dynamic, and mechanical signals all play a role. Materials such as ceramics, metals, fibers, polymers, and polymer composites may all be used to construct such scaffolds [6]. Scaffolds rely heavily on polymers because to these materials' capacity to mimic the organic matrix of healthy tissue while also being highly biodegradable and chemically inert (ECM).

Furthermore, the 4D printing method is used to form smart polymers of synthetic, natural, or hybrid origin into shape memory hydrogels and stimuli-responsive materials, resulting in programmable dynamical constructs with functional flexibility and shape regulation.

## 11.2 Next Generation of Bioprinting

Using bio-ink functional materials, 3D bioprinting technology can produce complex 3D tissue structures including cells to resemble genuine tissues. With this method, researchers have been able to create synthetic versions of skin, cartilage, and bone. Three main methods are used in bioprinting: extrusion, laser, and inkjet. To achieve 4D bioprinting, however, 3D bioprinted materials must be able to deform in response to external stimuli [7]. The final form of a 3D-printed item often undergoes shape-shifting postprinting in order to meet its functional specifications. The tissues may be replicated in a controlled and accurate manner. Moreover, it facilitates dynamic interaction with native cells to a limited extent. Understanding the distinction between 4D bioprinting and 3D bioprinting methods.

Dynamic 3D-printed biological constructions are created by the integration of 4D bioprinting methods with next-generation stimuli-responsive materials. Statistics from a recent publication show that the number of papers about 4D bioprinting using smart polymers has increased dramatically in recent years. The following section details crucial bioactive smart materials used in 4D bioprinting [8].

### 11.2.1 Medicine Management

Drug delivery systems (DDSs), tissue engineering (TE), and wound restoration are just a few examples of the many biomedical fields where 4D bioprinting technology has shown great promise. With 3D bioprinting, it is challenging to make hollow structures and blood arteries [9]. To combat this issue, researchers have developed a method of 4D bioprinting that produces a flat biological structure that, when subjected to a signal, folds into the blood tube conduit. Similarly, 4D bioprinting enables controlled medication release by delivering pharmaceuticals to a specific place in response to external stimuli; this diagram represents the whole trajectory by means of bioprinting in four dimensions [10].

There are several types of stimuli-responsive materials (SMs), however they may be broadly categorized as either shape-changing materials (SCMs) or shape-memory materials (SMMs), or as materials that change in stiffness or phase when subjected to external stimuli. When exposed to a new environment, SCMs will temporarily transform into a new form before reverting back to their baseline state. Hence, the deformation behavior of the materials has previously been determined. To generate complicated geometries in printed objects, SMMs use a shape-morphing effect and a sequence of programming processes. One subset of this bigger category (SMPs) is shape memory alloys (SMAs), and another is shape memory polymers (SMPs) [11].

The field of 4D bioprinting has seen the application of a broad variety of SMs. 4D-printed materials have shown a wide range of smart features, including shape-memory, self-assembly, self-actuation, self-sensing, and self-healing. In response to external stimuli, shape-memory materials may revert to a predetermined geometric form. Folding and assembly into a predetermined shape is triggered by environmental cues in self-assembly. The self-actuating mechanism, on the other hand, does not rely on external inputs to cause automatic actuation in materials. External stimuli may be detected and evaluated by self-sensing behavior. The self-healing process is the spontaneous repair of damaged structures. Many fields of biomedicine make substantial use of self-healing hydrogels, including tissue engineering, wound healing systems, dental restoration, and 3D bioprinting. In addition, knowing how different materials perform is crucial for 4D bioprinting [12].

### 11.2.2 Varieties of Bioprinting Material

Materials for biomedical applications must be able to deform in response to external stimuli, and this is made possible by preprogramming prior to 3D bioprinting. One-way, two-way, and multi-way deformations are distinguished within these materials. Stimuli-responsive materials are used in fourth-dimensional bioprinting because their final shape may be precisely controlled by activating an external signal. The physical (such as heat, sound, light, magnetism, and electricity), the biological include chemical compounds, enzymes, glucose, and cellular traction energies (such as salt) may all be used as external stimuli (pH value, humidity, and biomolecules). Materials that are responsive to a variety of stimuli are discussed in detail below, with an emphasis on their potential use in biomedicine [13].

#### *11.2.2.1 Thermoresponsive Materials*

To generate this transition in shape, an alarm signal of temperature increases is employed on thermoresponsive materials, which are triggered by a shift in temperature on the inside. Poly(-caprolactone) dimethacrylate, polyurethane (PU), poly(lactic acid), poly(N-vinylcaprolactam) (PNVCL), poly(N-isopropylacrylamide), poly(ethylene glycol) (PEG), collagen, and gelatin are all examples of SMPs (PLA). These high temperature SMPs are frequently used in 4D printing because to their great printability and processability. As if that were not bad enough, these SMPs have a glass-to-glass transition temperature that is greater than their minimum safety temperatures [14].

Just a select few materials have the inherent qualities to manufacture bio-inks for 4D printing, despite the fact that many synthetic-based SMPs display thermo sensitive behavior. Therefore, the glass transition temperature of materials used in the biomedical industry must be lower than 58°C to ensure the survival of cells. It is also possible to use a mix of biopolymers and thermoresponsive macromolecules to bring about the desired effect. Thermoresponsive synthetic materials, like PNIPAM, have been the focus of considerable use in thermal energy recovery and storage (TERS) systems, and their development has been the focus of many studies [15]. In a similar vein, bioactive materials undergo deformation due to solubility changes and temperature when they contract or expand. For instance, temperature-responsive shape morphing effect-printable Biopolymers scaffolds made from acrylated propyl soybean oil (AESO). At 37°C, the temperature at which the human body functions normally, the scaffolds reverted to their former shape, as seen by the authors. Gel, another temperature-sensitive biopolymer, also experiences a reversible gel–sol transition upon thermal change. For printing the hierarchical designs, they employed Gel, which responds to temperature, and chitin, which responds to pH, with citrate ions serving as the second cross-linker [16]. The Gel/Chit-H+ hydrogel was shown to be effective in molding responses to external signals and stimuli (citrate ion). Results showed that at temperatures over 30°C, Gel was soluble, whereas at temperatures below 25°C, it displayed a gel-like condition. A smart valve that controls water flow was printed using a dynamic Alg/PNIPAm ionic covalent entanglement (ICE) gel ink, however its temperature resistance is dependent on the gel from which it was made. The valve would shut and restrict flow by 99 percent when exposed to hot water, but would reopen when the water's external temperature decreased. Using 4D-printed actuating hydrogel materials, researchers looked at repairing catastrophic skin deformity and burned damage to the skin using a conjunction of zinc and glucophage coordinated complexes and poly(ethylene F127-based functionalized hydrogel [17]. The different combinations developed demonstrated excellent suppress reactive oxygen species, strong antibacterial activity, promotes cell proliferation, collagen synthesis, and angiogenesis, all of which are necessary for the effective treatment of traumatic skin defect and burn fine lines and wrinkles. In addition, studies of a high temperature hydrogel based on PNIPAM and enhanced with CS and Al3+ indicate that sprayable Pluronic F127 may be used as a medication delivery technique in the therapeutic treatment of skin injury. An intriguing option for use as a smart window material, the synthesized hydrogel exhibits a strong volume phase change in response to temperature rises of 20 C. To lessen the intensity of the afternoon sunset, researchers tested the biocompatibility of hydrogels made from thermo-sensitive hydroxypropyl chitin, N-acetylglucosamine (NAG), and carboxymethyl chitosan in mouse progenitor cells and male Sprague–Dawley rats. When the temperature of the produced hydrogel was increased, a successful sol–gel transition occurred [18]. Moreover, rat pancreatic beta cells were transported using thermosensitive hydrogels injected into various areas of diabetic rat. 3D culturing of cells in hydrogels was shown to boost cell proliferation, postpone apoptosis, and facilitate insulin production. Hence, cell engineering experiments using pluronic F-127–based hydrogels integrating mesoporous bioactive glass nanoparticles (NPs) for bone tissue regeneration show potential for the diabetic patient's treatment using the proposed method. The generated nanocomposite membranes hydrogels have shown promise for noninvasive administrations and for sol–gel phase transitions at two temperatures, making them appropriate for application as an injecting system, as shown by studies of their thermo-responsive reactivity and embeddings [19].

### *11.2.2.2 Biocompatible Polymers Materials*

Research into the use of magnetic and electrical fields in relation to biocompatible polymers has also received considerable attention. Para- and ferro-magnetic nanoparticles (MNPs) are examples of magnetic properties materials that may be used as additives in DDS. These sensitive polymeric materials may have significant applications in the biological sector. The therapeutic action of these compounds can be carefully controlled, making them ideal for use with DDSs [20]. For example, a magneto-responsive DDS for the rehabilitation of damaged soft tissues might be made using an agar/PEG-based hydrogel and a network of Fe3O4-based NPs. Hydrogels with an excellent rheological characteristic make magneto-responsive 4D-printed bioproducts possible, and they have been made possible by reinforcing Fe-based NPs into poly(dimethylsiloxane) (PDMS). As magnetic bio-inks provide more exact control over the orientations of the NPs as they are printed, they may be useful in the biomaterials technique, which produces scaffolds with anisotropic characteristics [21].

Other examples of ferromagnetic biodegradable polymers developed for TE applications include electromagnetism scaffolds of Fe3O4/polyethylene glycol diacrylate (PEGDA) and Fe3O4/PCL, and the approval process of transversely isotropic Fe3O4 magnetic NPs with PEGDA hydrogel via a digital light processing (DLP)-based 3D printing process. Human mesenchymal embryonic stem cells (MSCs) demonstrated excellent cytocompatibility when these magnetic NPs were directly planted on them. They showed exceptional survival of more than 95% of the human skin fibroblasts when mixed into polymeric nanoparticles. Hence, unpaired movements are enabled after the magnetic NPs are aligned with the proper scaffold configurations. Although these magnetic NPs have many biological applications, TE is only one of them [22].

#### *11.2.2.3 Endophyte Biocompatible Polymers Materials*

Electro-responsive products offer use of biomaterials like polymers that can conduct electricity. Bioelectric stimulating of endophyte biocompatible polymers allows for the reversal of constriction, contraction, or folding by adjusting the intensity of the electromagnetic field and the external environment [23]. Sometimes, the dielectric properties of these material are improved by adding additives like carbon nanotubes (CNTs), graphene, and other NPs. For instance, a 4D-printed origami robots used a poly role-based organic polymer that was electro-sensitive. Researchers are now investigating the use of these biocompatible polymers to develop muscle and neurons TE constructs. Because of their superior printing capability and cytocompatibility, hydrogels based on polymer nanocomposites may also be employed in composite structures of biomolecules [24].

#### *11.2.2.4 Photo-Conductive Polymer Materials*

In order to bend, compress, or otherwise change the volume or size of the structures, photo-responsive materials may be used as a stimulus. Photoconductive nanostructures include polymers incorporating stilbene as well as derivative of aromatic hydrocarbons, cause of pollution, fulgide, and spiropyran. By incorporating chromophores into polymer resins, photo-responsive polymeric materials may be created [25]. The incorporation of photosensitive nanomaterials into polymers offers a similar avenue for material development. The system's reaction may be permanent or temporary depending on the chromophore. Several of the polymeric materials used in DDS development are irreversible photo-responsive compounds known as photodegradable polymers. Insightful investigation is necessary for TE applications when Bioprinting in four dimensions using photo-activated polymers that change shape when exposed to light. Recently, novel illumination materials were developed [26] by surface functionalization PNIPAM hydrogels with spirobenzopyran.

### *11.2.2.5 UV-Assisted in 3D Printing*

In the humid environment or cell culture medium, moisture may be absorbed or lost by materials that react to humidity by swelling, folding, twisting, or deforming. These SMs may find utility in actuating and soft robotics applications due to their ability to swell and contract. For instance, stearoyl moieties were added to microcrystalline cellulose (MCC) to make a water-reactive substance [27]. Researchers have developed a 4D-printed material that is flexible and can change shape in reaction to the moisture present using UV-assisted direct-ink-writing-based 3D printing. When exposed to moisture, the printed structures were distorted, however drying out returned them to normal. Further research into this characteristic of diverse biopolymers might lead to the development of novel biomedically useful materials. Many other polymeric polymers, including cellulose, silk fibroin, PU, and PEGDA [28], employ a similar desorption or sorption of humidity as the actuating response force. Researchers examining the responsiveness of silk fibroin films to the presence of ethanol and water vapor found that the films underwent reversible, multiple unidirectional, macroscopic scales actuation. When water is present, the thickness of the film may fluctuate between expansion and contraction only little, allowing for huge, reversible activation. The actuator may react to predetermined stimuli such as liquid water and ethanol vapor, allowing it to undergo a three-dimensional shape change. The suggested technique reinforced a hydrogel by inserting tunicate cellulose nanocrystals (TCNCs) into polymeric networks through mediated activation, emulating the function of natural tendril muscles [29]. In the future, this may be used in hydrodynamic technologies used in labs on a chip to pump fluids. The hydrogel muscles' actuating rate, activation strain, and shape retention all significantly increased in response to stimuli. It was found that the chirality, pressure, twist density, and transitory shape of hydrogel muscles all have a role in their ability to actuate. Because of their high percentage of water and stretchy work capability, manufactured hydrogels provide such a lot of opportunities for the field of biomedicine [30].

### *11.2.2.6 Sensitivity Polymeric Materials*

Materials having functioning reactive groups, such as pyridine, carboxyl, sulfonic phosphorous, or acid, are pH-responsive biomaterials. As the ambient pH changes, certain chemical groups in biomaterials either lose or gain protons, causing the materials to contract or expand, respectively. Basic polymers (those with an NH2 group) and acidic polymeric are the two main categories intricate tissue development, including the incorporation of vascularization functions, will benefit from the production of printed hydrogels of increasing complexity. Gene delivery systems, drug delivery systems, and glucose sensors are just a few examples of the many biomedical applications that have taken use of pH-sensitive materials and their unique features. The sensitivity of these polymeric materials makes them useful in TE applications, and the varying pH levels in various human body sections may be taken advantage.

### *11.2.2.7 Macromolecules Materials*

Bio-enzymes, glucose, and biomolecules are only some of the biological stimuli that might trigger these materials. An intriguingly applicable idea suggests that some bio-enzymes may break protein sequence and macromolecules, generating alterations in the polymers that enable them to grow. Several TERM-related uses may benefit from this shape-shifting effect. For instance, thrombin and alkaline phosphatase were used to activate hydrogels that were designed to mimic biological tissue. This caused calcification and fibrin accumulation on the 4D-printed object. Fourth-dimensional printing, which responds to biological cues, may also be used to create the functionalized hydrogels. In order to achieve 4D printing, it is necessary to investigate the interplay between biological reactions and other physical or chemical responses.

### *11.2.2.8 Dual-Sensitive Materials*

Certain substances respond best to a mixture of stimuli, whereas others need a single stimulus. Due to their complex and varied reactions, these materials are much admired in the scientific world. Just as TERM, DDS, and soft robotics may all benefit from materials that can undergo chemical and structural changes, so can multi-sensitive polymeric materials. For instance, newly developed polymers are based on the potential to be used in the biomedical area because of their sensitivity to both pH and temperature. hydrogels, which are pH and temperature sensitive dual-responsive hydrogels for cardiac tissues, were also synthesized via free radical copolymerization. Research on hydrogels has shown solidification at 37°C. They discovered that cells grown from the cardiosphere, the heart's topmost layer, may survive in a specialized capsule for up to a week.

Similarly, Fe3O4-based NPs were incorporated into a microgripper using a thermomagneto-sensitive polymer. Adding azobenzene chromophore to N-isopropylacrylamide thermoresponsive polymeric polymers makes them both transparent and thermally. Polymeric materials that are responsive to physical, physiological, and physiological inputs have also been produced, and there is an exciting work that incorporates graphene oxide (GO) into a PNIPAM-based hydrogel to generate dual-sensitive materials that might be employed in DDS. The polymers' adaptability suggests several potential applications in the biomedical sector, including tissue engineering (TE), medication delivery, and bio-adhesives.

Triple shape memory effect (TSME) of metallopolymer networks was shown when two transitory forms were exchanged, activating and deactivating metal-ligand interaction with differing stabilities. Recovery rates for the Fe-crosslinked polymer were reported to be extraordinarily high. Ternary shape-memory blends of thermoplastic urethane (TPU), polylactic acid (PLA), and poly (propylene carbonate) (PPC) were developed, with Fe/Zn and Co/Zn polymer exhibiting the highest strain fixity rates during the first fixation phase. The dissimilarities in compliance, surface energy, and viscosity between TPU and the compound led to the development of the novel co-continuous structure. Moreover, the concentration of LPC may be adjusted to tune the surface energy between the compounds and TPU, resulting in adjustable dual and TSME. The TSME of zwitterionic PAEK was investigated by creating a blend of PL-prepared structure PAEK and grafted carboxylic acid pyridine zwitterions. The phase shape and thermal-mechanical characteristics of this mixture allowed it to perform very well in TSME. Specifically, it was revealed that interactions between molecules of water and carboxylic acid pyridine zwitterions were responsible for the observed rearrangement behaviors. In addition, the sample's high TSME was achieved by its capacity to take in water. Moreover, the temporary shape-based sample was restored to its original shape after two stages and the planned form fixing processes. This zwitterionic PAEK TSME can detect water and change form in response to it. It was also investigated how the interaction between poly and iron oxide (Fe3O4) affected the shape-shifting behavior of the multi-material. These diverse materials were sensitive to magnetic stimulation thanks to the incorporation of Fe3O4 NPs. When subjected to combined thermostimulation and magnetic stimulation, the multiple-SMP and its nanocomposites exhibited excellent shape-changing behavior with more freedom of motion in the metamorphosis than originally explained.

### 11.2.3 Biosensing Scaffolds

The use of SMPs in biological devices like cardiac patches, stents, cardiorespiratory implants, orthopedic surgery augmentations, electric motors, skin condiments, biosensing, scaffolds, and tourniquets has skyrocketed in recent years due to their superior permanent deformation feature compared to SMAs and their demonstrated great with modern manufacturing techniques like 4D printing. This means that the polymers may be modified to suit the needs of certain applications and patients. Recent studies have focused on the 4D bio-printed SMP products' *in-vitro* and *in-vivo* responses. Polymer composites and hydrogels, printing resolution, and internal as well as external stimuli all play a role in the evolution of 4D-printed products with specific biomedical uses.

Citrus pectin/beta-cyclodextrin (-CD) complexes were investigated for their potential to stabilize a pickering emulsion for use in four-dimensional printing. The pH-sensitive curcumin added to the emulsion made it possible to utilize in 4D printing. All of this occurs in the setting of the proposed origami-designed-based stents with self-assembly via 4D printing; heating causes the dissolution of different ions like $HCO_3$ into CO, leading to higher pH values; published samples start changing from yellowish to red; and the chroma of the specimens changed depending on various dimensional values. The shrinkage ratios of a number of stents are calculated, all of which have different helix angles and unit cells. The results indicated that helical angles improved the shrinking ratio of the stents. Lightweight design also improves structural flexibility, reduces shrinkage ratio, and adjusts the location of the flexure hinge. Heat-induced self-assembly is shown, and the stent's folded and expanded states are contrasted to highlight its potential in future tissue engineering applications. Using holographically shifted femtosecond multifoci, a pH-responsive hydrogel was fabricated for use in 3D functional microarchitectures. The engineered microcage array, activated by pH shifts, demonstrated very efficient target capture. Moreover, the engineered scaffolds served as arrayed analytical platforms for monitoring cell patterns in static or moving environments and for showcasing the anticancer efficacy of drug payloads. Therefore, functional microstructures based on hydrogels for manipulating micro-objects may be made using the proposed methods. Particle jamming and extrusion-based 3D printing were investigated as potential methods for fabricating porous microgel scaffolds from PEG-based synthetic microgels with variable degrees of degradability. Large-scale 3D culture of cells in a macroporous hydrogel was also demonstrated by high-throughput direct encapsulation of cells in printed microgels. As a result of the unprecedented spatiotemporal control over the features of printed microporous annealed particle, printed scaffolds are now a promising material for 2D and 3D tissue culture.

An anisotropic stimulus response was produced by the films because of the inhomogeneous illumination experienced during projector-based photopolymerization. Moreover, a helical coil, cone, handedness-inverted coil, twisted flower, or saddle may form if the crystal orientation is patterned spatially. The temperature at which the actuator is triggered may be controlled by adjusting the concentration of the chemicals utilized. The proposed approach offers significant potential for a wide range of applications, from deployable electronics to morphing biomedical devices, due to the volatilization of a non-fully-reacted, volatile component in a partially cured cross-linking network generated by photopolymerization. Several sophisticated shape changes, such as the self-folding of a buckyball from a 2D hexagonal lattice sheet or the expansion of numerous pop-up structures from their original compact configurations, were seen due to volume shrinkage induced by the loss of the volatile component.

The miura-ori structure is an example of smart materials that can sustain over number of times its own weight and is made by integrating metal coordination bonds with shape memory PAEK. Similar to the complex formation of the intrinsic movement potential of the PAEK matrix and the cracking- recombination of strong interaction, tuning the metal ion content and species led to the excellent polymeric behavior observed in terms of high recovery ratio. The dynamic cracking-recombination of coordination bonds provided novel 3D metamaterials with zero Poisson's ratio for reversible energy absorption applications, which gave PAEK material its extraordinary self-healing and reprocessing capabilities. Metamaterials' outstanding mechanical capabilities were also shown by thermal testing through a heating-cooling process, during which plastic deformations were completely released and samples resumed their prior shapes.

## 11.3 Biosensors and Actuators

The materials used in 4D printing need to print well because of their superior shape memory characteristics, biocompatibility, and simplicity of processing, SMPs are among the most crucial smart biomaterials used for 4D bioprinting. This section will explain the most current 4D bioprinting applications, which include the fabrication of various SMs, including as biosensors and actuators, for use in biomedical settings.

Scaffolds: Intricate tissues can be developed using 4D bioprinting technology, which can also alter the form and function of polymeric materials to create the necessary cellular stresses. Scaffolds, which may help in the rebuilding of injured tissue, can be manufactured using this technique, which has lately demonstrated significant promise. Medical applications benefit greatly from the excellent *in-vivo* and *in-vitro* capabilities of compostable polymer-based scaffolding having a porous structure. Shape modification is the most common use for scaffolds like this, but they may also be used to modify both their electrical and mechanical characteristics. Scaffolds with a greater shape memory effect may be created using plant-based biopolymers, in particular AESO. After suffering a transient geometry shift at 18 C, it was discovered that an AESO-based oil polymer restored to its permanent form at normal temperature. *In vitro*, investigations showed that human migratory cancer stem cells were capable of self-renewal, adhered well to other cells, and could develop into several cell types while retaining their functional ability (hMSCs). To apply mechanical stimulation to regulate cells, similar initiatives have employed porous scaffolds made from PU-based SMP. At human body temperature, the findings showed that the scaffold had a great form morphing action and returned to its previous shape. These 4D-printed scaffolds may be implanted by non-invasive methods.

Biopolymers such as lignin and cellulose also react strongly to changes in humidity. For instance, by experimenting with different proportions of lignin and keratin, 4D-printed goods have been generated. Unfortunately, the findings also revealed that filaments based on lignin and keratin degraded when exposed to high temperatures during fabrication. To address this problem, researchers employed paste printing techniques, whereby hydrogels with 30%-40% solid content were created and shown outstanding characteristics, guaranteeing their usage in 4D-printed scaffolds. Hydrogels made from the biopolymer Alg/HA were used in another investigation. Mice mesenchymal stem cells were included in 4D-printed self-folding structures. After 7 days, the findings showed that the cells were still alive and well, indicating high biocompatibility. Scaffolds made from hydrogels based on biopolymers were also shown to be feasible according to the findings of this research. Excellent shape morphing effects following temperature stimuli were also seen in a hydrogel created from a CS-based biopolymer using the SLA approach, suggesting its potential use in the generation of scaffolds.

### 11.3.1 Fabricated Scaffold Tissues

Bones have the ability to self-heal and mend minor breaks and flaws. 4D bioprinting aids the 3D-printed scaffolds in reshaping in response to environmental stimuli, allowing for more effective bone grating in the context of these bone deficiencies with uneven geometries. Four-dimensional bioprinting is an intriguing method for creating Integrated osteogenic peptides and black phosphorus nanosheets into photothermally sensitive tricalcium phosphate. The results showed that the morphology scaffolds may be modified by the activation of light-sensitive BPNSs using near-infrared radiation (NIR). When the scaffolds were heated to core temperature, they had mechanical qualities similar to the cortical bone of the individual. *In-vivo* studies found that when these scaffolds were used to treat cranial abnormalities in rats, the animals' bones healed more quickly.

To help in the transport of oxygen and nutrients into the scaffold's core, swelling polymer composites is another method for solving the bone grafting issue. Hydrogel composite bio-ink, derived from plants and made of acrylamide matrices and cellulose fibrils, was utilized to make a biomimetic 4D-printed item with working parts. Cellulose fibers were aligned in a single direction throughout this printing process, leading to an increase in thickness along the length of the object. Likewise, cell-laden scaffold tissues were fabricated The study confirmed that cell-rich scaffolds might differentiate into bone-like tissues.

The success of 4D bioprinting in human and animal subjects is crucial to the technology's widespread adoption. One of the *in-vivo* studies is that conducted who employed HA/PCL-based SMP to develop porous scaffolds for transplantation in mandibular fracture healing for BTE. These results demonstrated that SMP porous scaffolds may be deformed into non-porous morphologies under pressure, but would return to their original shape when brought up to body temperature. Despite the significant amounts of heterogeneity present *in vivo*, these smart 4D-printed scaffolds show a great deal of potential.

Amazing photothermal therapy (PTT) for tongue cancer was made feasible by developing an injectable hydrogel that responds to near-infrared light. These synthetic fibers are very biodegradable and show strong photothermal effect since proteins, Ag3AuS2 NPs, and CS molecules are all negatively charged. Results from a mouse model of in-the-mouth tongue cancer showed that a single dose of PTT was enough to eradicate tumors without harming adjacent tissues. This bodes well for potential future therapeutic use in the elimination of oral cancers. Vascularization and skin flap regeneration may benefit from the MX-HF scaffolds' continuously responsive behavior caused by the photothermal conversion capacity of the MXene nanosheets and the thermally capability of poly (NIPAM) hydrogels. Cells were able to easily invade the scaffold channels from the outside thanks to their NIR-responsive water retention activity. By increasing angiogenesis, decreasing inflammation, and protecting against apoptosis, the NIR-responsive scaffolds greatly increased skin flap survival *in vivo*. This makes MX-HF scaffolds applicable in clinical settings where regenerating random skin flaps is required. The easily available terpenes limonene and myrcene were used to develop oligomeric photopolymers for DLP-based 3D printing. The manufactured photosets displayed 4D shape memory capabilities and tuneable thermomechanical properties. These photopolymers have biological uses as scaffolds.

### 11.3.2 Vascularizing Tissues

Vasculature models of different sized have been created using 4D bioprinting, exhibiting the flexibility of this approach to solving the difficulties of vascularizing tissues. This is made possible by the self-folding framework that converts the 2D-planar geometry to the 3D micro-scaled hollow tubules (MHTs) in reaction to an external stimulation. These models experience localized remodeling thanks to the use of bio-ink materials with time-dependent alterations in physical properties.

In contrast to 3D-bioprinting, 4D-bioprinting facilitates the creation of MHTs with consistent cell dispersion. These days, SMPs, photo cross-linkable hydrogels, and MSCs are the materials of choice for 4D bioprinting projects involving the creation of self-folding structures. Microvascular scaffolds may be built using water stimulus 4D-bioprinted GelMA/poly(D,L-Lactic Acid) (PDLLA)-co-TMC by adjusting the degree of alignment of the hydrogels. The microvascular scaffolds generated folded and had internal diameters comparable to blood vessels when heated at a high temperatures. These MHTs were also superior in terms of biocompatibility and cellular proliferation. In addition, extrusion-based bioprinting technique has been used to create molten electro-writing polymer nanocomposites and photo cross-linkable hydrogels. Just recently has 4D printing been used to create vascular models. The direction of cells in vascular networks and the controllability of the strain in 4D-printed structures in response to external stimuli are only two examples of the many issues that need to be addressed.

For myocardial repair, researchers have employed a 4D-printed cardiac scaffold made of polydopamine (PDA)/alg and a cell-laden hydrogel triggered by NIR. Cells cultured on these scaffolds for two weeks demonstrated remarkable viability despite their abnormal shape. In a similar vein, we employed SLA technology to create a cellular patch comprised of human induced pluripotent stem cell-derived cardiomyocytes (hiPSC-CMs), human endothelial cells (hECs), and human mesenchymal stem cells (hMSCs). Myocardial infarction therapy has also made use of myofibrils with different orientations (MI). Based on the results of *in vivo* testing, the cellular 4D-printed patch exhibited improved cytocompatibility, blood density, and heart adhesion. HiPSC-CM-based tissues were produced using ultrasound. These 3D-printed tissues improved cardiac rhythm, contractile stress, and relaxation-contraction rates. This promising finding has implications for the development of 4D bioprinting.

Instead of using these synthetic polymers, researchers have found significant success using a plant-derived polymer called AESO, which exhibits better biomechanical efficiency and excellent adherence to wounded cardiac tissues. Methacrylate, hyaluronic acid, and alginate-based hydrogels are used to print mesenchymal hematopoietic tissue (MHT) that may serve as blood vessels. Research on the potential of 4D bioprinting for cardiac TE applications is ongoing, but the field still requires advances in bio-ink materials and cellular patterning. As these challenges are resolved, 4D-bioprinted heart elements may help patients with cardiovascular problems.

### 11.3.3 4D Bioprinting Neural Tissue

In neural tissue engineering (NTE), scaffolds that match the form and composition of extracellular matrix (ECM) may improve cellular adhesion, proliferation, and differentiation. 4D-printed micro-grooves aid in the reconnection of nerve fibers, which is crucial for nerve function. 4D bioprinting employing bio-ink incorporating AESO, hMSCs, and graphene has been employed in the NTE technique to generate nerve guidance conduits. The photo-sensitive AESO-based polymer caused the printed structure to bend, and the inclusion of graphene-based NPs increased the biomaterial's permeability and helped hMSCs differentiate into brain cells. When AESO-based SMP reverted to their permanent form at 37 °C, researchers were able to design smart NGCs with functionalities such as dynamic tubulation, chemical cues, physical guidance, and seamless integration, all of which are important in nerve transplantation.

Integrating dispersed neural stem cells (NSCs) into thermosensitive PU-based hydrogel prior to the inclusion of gelatin increases the stiffness properties of the hydrogel without the necessity of a crosslinker. The results showed that at 37°C, bioprints multiplied and discriminated well.

Successful nerve repair in a zebrafish model using a biodegradable hydrogel created on a 4D printer. A poly(glycerol sebacate) based bilayer mat was also used to produce NGCs. The results demonstrated that printed soft NGCs have excellent biocompatibility and degradation resistance. Neurons were able to successfully adhere, differentiate, and proliferate on artificial nerve grafts. Animal studies of neuron regeneration demonstrated the potential of a self-healing hydrogel created via 4D bioprinting.

### 11.3.4 Longitudinal Deformation

In order to repair injured cartilage, the 4D bioprinting process may be used to make scaffolds for cartilage tissue engineering (CTE). Using 4D bioprinting technology, researchers have developed curved trachea implants for cartilage healing. Nanocellulose fibrils that are arranged in a straight line within acrylamide matrices experience longitudinal dislocation when they are hydrated. The expanding process allows for even the most complex curves to be filled. A trachea implant, for instance, might be printed out of a photo-crosslinked silk fibroin-based hydrogel using a 4D printer. Researchers stuffed a hydrogel structure with cells using digital light processing technology, and then implanted it into rabbits to evaluate the *in-vivo* response. According to the results, rabbit tracheas repaired cartilage effectively.

We considered CTE applications while we designed our synthetic fibrocartilage and employed in-printing matrix remodeling to steer collagen fibers in the right direction. The authors were able to develop magnetic chondrocyte-loaded bio-ink tissue scaffolds with many layers of scaffolding. Engineers have developed functionally graded polymeric materials with adaptive mechanical characteristics via 4D printing. These materials were inspired by natural phenomena such as fish scales and tendon-bone attachment. The authors claim that printing the supple muscles that surround the skeleton is possible with these materials. Methods of Medication Dispensing

With the use of 4D bioprinting technology, a wide range of DDSs, including as tablets, multilayered DDSs, transdermal systems, orodispersible films, dermal skin patches, nanosuspensions, rectal and vaginal delivery systems, and more, may be manufactured with unrivaled accuracy. By responding to external inputs, 4D bioprinting might help with the transport of bio-molecules in addition to enabling for fine-tuning of mechanical strength and shape. Stimuli-responsive drug delivery systems (DDS) have gained popularity as a means of better controlling medication delivery characteristics. Recently, pH-sensitive sensors have been included into smart automated bandages, which enables the release of medication in response to signals, stimuli, or abnormal clinical situations.

Drug release control with the use of 4D-bioprinted DDS with a stimulus-dependent response. To treat the digestive tract, for instance, researchers developed a thermoresponsive drug delivery system (DDS) based on PNIPAM-AAc/poly(propylene fumarate) (PPF). Helical micro-swimmers for targeted DDS have been created by integrating MNPs into the photocurable resins. A CS-based UV-activated helical micro-swimmer was developed by the authors of a different study. The results indicated that the micro-swimmer that was created had excellent good biocompatibility and drug-releasing ability. In a similar vein, we designed a 4D-printed capsule with a core-shell hydrogel system to regulate the slow release of the drugs. The capsule's inside was constructed of ethylene glycol, poly(vinyl alcohol), and bio-molecules, while its outside was built of gold nanorods (AuNRs) and poly(lactic-co-glycolic acid) (PLGA). These hypersensitive AuNRs, which carried pharmaceuticals, burst when triggered by a laser and released their contents.

Multistimuli-responsive triblock copolypeptides with fast and regulated drug release were developed by combining thermosensitized elastin-based polypeptides (EBP) with ligand-responsive calmodulin. The thermally induced gelation of concentrated ECE triblock copolypeptides led to hydrogels with physical crosslinks under the same conditions. Because of the shown conformational change of the calmodulin middle block following binding of and trifluoperazines ligands, these hydrogels exhibited controlled mechanical and rheological properties and high biocompatibility. In the biomedical field, the stimuli-responsive ECE triblock hydrogels might be used as injectable drug delivery depots. Microrobots based on a pH-sensitive hydrogel with the ability to change form in response to their environment were developed (SMMRs). Several rates of expansion were programmed into the SMMRs that were generated. To protect the drug doxorubicin (DOX), a shape-shifting microfish (SMMF) was designed to seal its mouth in phosphate-buffered saline (PBS) and open it in a slightly acidic solution. Moreover, human cervical cancer cells were treated locally using an artificial vascular network activated by the "opening closing" of the SMMF mouth. So, it is great for a variety of uses, such as the controlled delivery of medications and the transfer of intricate microcargo.

Using a magnetorheological fluid, it demonstrated a magneto-active based millirobot with the ability to rapidly and easily transition between a solid liquid state at low magnetic fields (0 mT) and the Bingham plasticity in the solid state at high magnetic fields. When in a liquid state, the developed MRF robot is able to perform a wide range of tasks, including large deformation, *in situ* splitting, smooth navigation, merging, and gradient pulling. When in a crystalline structure, the robot is able to perform a wide range of tasks, including easily controlled locomotion with reconfigured shapes and a broad range of object manipulation techniques (including push, pull, and rotation of the objects). In this work, we employ magnetic actuation to show that the MRF-Robot can be used to effectively deliver liquid medications, remove thrombi, and restrict fluid flow inside a phantom vascular model. Domestic Medical Technology

Medical equipment such as stents, vascular repair devices, and orthopedic implants have grown increasingly complicated and personalized to the individual with the advent of 4D bioprinting. Due to its ability to precisely control medication release, SMP-based drug-eluting stents have found utility in medicine lately. These stents are also often used for vascularization. These stents not only dilate blood vessels, but also help disperse anti-platelet and anti-proliferation drugs. Perhaps less invasive procedures might benefit from 4D-printed stents. Because of their superior biomechanical resistance, helical products are often made using thermoresponsive and chemo-triggered polymeric materials. Platelet adhesion, biocompatibility, and efficient release are also considered essential features.

It investigated 3D microfluidic devices that can be programmed for different shapes using a bilayer of channel-embedded PDMS and SMPs, and it did so for a wide range of geometries, including open-mesh topologies. The variety of 3D structures available for use in microfluidic devices has expanded considerably. The programmed behavior of shape morphing was reversed in response to temperature inputs. It was shown that fluid flow was maintained successfully in both the deformed and repaired microfluidic channels. Furthermore, 3D microfluidic devices were rapidly actuated by adding magnetic particles to the PDMS layer and manipulating them with a hand-held magnet. As a result of these advancements in open-mesh 3D microfluidic devices with completely programmable behavior, biomedical applications ranging from medication delivery to tissue engineering have been realized. In a different study, functional biorobots were developed by combining live cardiomyocytes with tunable sound fields. Controlled actuation of a soft skeletal system and injecting of microcapsules, as well as through found naturally contraction-relaxation periods, are just a few examples of how these biorobots have implemented dynamic thiocarbamate bonds into the photocurable methacrylate to create reprocessable and self-healing polythiourethane (PTU)-based different objects. Form fixity, shape recovery, surface wettability, and biocompatibility were all shown by printing a hand model with the shape-memorized PTU.

Especially in the realm of bio-implants, the 4D-printed PTU offers a wide range of applications.

The effect of biodegradability on the mechanical qualities of micro-stents is also considered throughout their development. Recently, 4D printing technology was utilized to construct high-resolution SMP-based thermoresponsive cardiovascular stents, and FDM technology was used to fabricate a polymer-based cardiovascular stent, both of which are used to expand blood vessels and enlarge tracheas. The constructed stent lost its form and developed fractures when heated to 37°C. The *in-vivo* reaction, however, requires more study. A high-tech stent based on PNIPAM-AAc/dipentaerythritol pentaacrylate (DPEPA) was also developed using DIW technology. The finite element method (FEM) was utilized to refine the design depending on the response to stimuli. It was shown that the novel 4D-printed stent responded to changes in both temperature and pH. A strong relationship was also found between the stimuli-response and the material's stiffness. Such tools may be used to create goods that can keep up with the dynamic nature of today's workplaces.

A tracheal-bronchial stent is a hollow tube used to widen the airways and help a patient breathe. When faced to the continual fatigue strain of breathing, traditional trachea stents remain in the same place. Hence, employing 4D bioprinting technology, stimuli-responsive stress-free personalized stents have been produced. By sensing their surroundings, 4D-printed morphing stents can respond to the dynamic stresses induced by the body's temperature. Bioresorbable tracheal stents made from SMPs manufactured in 4D using SLA technology. The 4D-bioprinted tracheal stent changed itself to fit its new environment, as shown by *in vivo* data. Similar shape memory behavior was seen in a 3D printed Fe3O4/PLA tracheal stent when exposed to magnetic stimuli and UV radiation. This stent caused little harm in terms of both volume and severity. The 4D bioprinting technique has been used to effectively create artificial scaffolding for intravascular and tracheal stents.

### 11.3.5 Uses of Biomedical Appliances

Bionic limbs, splints, and elbow guards are just a few examples of the many biomedical uses that might benefit from 4D bioprinting. Another groundbreaking study suggested radioscopic analysis of 4D-printed porous PU-based SMP. When salt sodium chloride was added, porosity occurred, and when tungsten was added, radiopacity was achieved. After being rinsed with distilled water, the resulting biomaterial, an endovascular coil may be placed over a radiopaque porous 4D-printed object. Extreme measures are being taken to design and print gland tissues utilizing 4D bioprinting. It is a major challenge in oncology to recreate the tumor microenvironment *in vitro*. A PEGDA/GelMA-based bi-layered hydrogel was made and then seeded with cells. These biocompatible, curved, tubular, and foldable structures are essential for testing ductal cancer *in vitro* models. Technology for 4D printing is now being used at the cellular level. Using an oil droplet printing technique, a water-based tissue-like structure was created. They used light stimulation to replicate functional neural transmission after printing cell-free expression into an oily environment.

## 11.4 Enhancing Technology in Bioprinting

Bioprinting is a cutting-edge technology that has the potential to revolutionize fields such as regenerative medicine, tissue engineering, and drug discovery. It involves the precise deposition of biomaterials, cells, and other bioactive components in a layer-by-layer manner to create functional three-dimensional structures.

To enhance technology in bioprinting, researchers and engineers are focusing on several key areas:

Biomaterials Development: Improving the bioinks used in bioprinting is crucial. Scientists are working on developing bioinks that mimic the properties of native tissues, including mechanical strength, biocompatibility, and degradation rates. This involves exploring novel materials such as hydrogels, synthetic polymers, and decellularized extracellular matrices.

Cell Sourcing and Characterization: Bioprinting relies on viable and functional cells. Advances in cell sourcing, isolation, and characterization techniques are essential for obtaining high-quality cell populations for bioprinting. This includes optimizing cell culture conditions, using induced pluripotent stem cells (iPSCs), and exploring cell reprogramming techniques.

Printing Techniques: Various printing techniques are employed in bioprinting, including inkjet, extrusion, and laser-assisted methods. Enhancements in these techniques aim to improve printing resolution, speed, and accuracy. For example, the development of nozzle designs, control systems, and bioprinter hardware/software integration can contribute to more precise and efficient bioprinting.

Vascularization: One of the significant challenges in bioprinting large and complex tissues is the integration of a functional vascular network. Researchers are exploring strategies to bioprint blood vessels and capillary networks within engineered tissues. This involves the development of bioinks containing endothelial cells, angiogenic factors, and sacrificial materials that can be removed to create hollow channels.

Bioreactor Systems: Bioprinted constructs require a suitable environment for cell maturation, tissue development, and functionality. Bioreactor systems that provide appropriate mechanical and biochemical cues to the printed tissues are being developed. These systems enable the integration of perfusion, mechanical stimulation, and biochemical gradients, which are crucial for tissue growth and maturation.

Multi-material and Multi-cellular Printing: Advancements in bioprinting aim to enable the simultaneous deposition of multiple biomaterials and different types of cells. This capability allows the creation of complex, heterogeneous tissues and organs. Integrating multiple bioinks, optimizing printing parameters, and precise control of cell distribution are key areas of research.

Biofabrication Standards and Regulations: As bioprinting progresses towards clinical applications, the establishment of standards and regulations becomes vital. This includes defining quality control measures, safety guidelines, and ethical considerations to ensure the reproducibility, reliability, and safety of bioprinted products.

Computational Modeling and Design: Computational modeling techniques play a crucial role in optimizing bioprinting processes and predicting tissue behavior. By combining mathematical modeling, computer simulations, and experimental data, researchers can improve the design and fabrication of bioprinted structures.

## 11.5 Conclusion and Future Work

In conclusion, the integration of cyborg intelligence and bioprinting holds tremendous promise for the computational design and analysis of medical applications. By combining the power of artificial intelligence, advanced computational modeling, and bioprinting technology, we can create innovative solutions for personalized medicine, tissue engineering, and regenerative therapies. Cyborg intelligence, which refers to the seamless integration of artificial intelligence and human capabilities, can enhance the design process by providing sophisticated algorithms, data analysis, and decision-making tools. It enables the optimization of bioprinting parameters, the prediction of tissue behavior, and the generation of personalized treatment plans. Bioprinting, on the other hand, allows for the precise fabrication of complex three-dimensional structures using biomaterials and living cells. It enables the creation of functional tissues, organoids, and even organs, which can be used for drug screening, disease modeling, and transplantation.

By combining these two technologies, researchers can harness the power of computational design and analysis to drive advancements in medical applications. Some potential future directions include:

Personalized Medicine: Cyborg intelligence and bioprinting can be integrated to develop patient-specific treatments. By analyzing medical data, genetic information, and computational models, personalized therapies can be designed and bioprinted to match the individual needs of patients.

Organ-on-a-Chip Systems: Bioprinting can be combined with microfluidics and cyborg intelligence to create organ-on-a-chip systems. These platforms mimic the structure and function of human organs, enabling high-throughput drug testing, disease modeling, and toxicity studies.

Artificial Organs and Tissues: Advances in bioprinting techniques and computational modeling can lead to the fabrication of functional artificial organs and tissues. These constructs can be used for organ transplantation, reducing the dependency on organ donors and addressing the organ shortage crisis.

Simulation and Virtual Testing: Cyborg intelligence can be employed to develop advanced computational models that simulate the behavior of bio-printed tissues and organs. Virtual testing can help optimize the design parameters, predict the performance of bioprinted constructs, and reduce the need for extensive experimentation.

Regulatory and Ethical Considerations: As these technologies advance, it is essential to establish regulatory frameworks and ethical guidelines. Future work should focus on addressing safety concerns, quality control, and ensuring the responsible and ethical use of cyborg intelligence and bioprinting in medical applications.

## References


1. Yu, Y. and Chiu, I., Cyborg intelligence: Recent advancements and future perspectives. *Adv. Intell. Syst.*, 2, 12, 2000080, 2020. https://doi.org/10.1002/aisy.202000080.
2. Tao, J. and Zhang, Y.S., Bioprinting for neural tissue regeneration. *Adv. Healthc. Mater.*, 9, 13, 1901647, 2020.
3. Martínez-Moro, E., Navarro, M., Guillem-Marti, J., Samitier, J., Serena-Pou, J.M., Bioprinting techniques: From design to fabrication for tissue engineering applications. *Materials*, 13, 8, 1777, 2020.
4. Murphy, S.V. and Atala, A., 3D bioprinting of tissues and organs. *Nat. Biotechnol.*, 32, 8, 773–785, 2014. https://doi.org/10.1038/nbt.2958.
5. Kaswan, K.S., Dhatterwal, J.S., Balyan, A., Intelligent agents based integration of machine learning and case base reasoning system, in: *2022 2nd International Conference on Advance Computing and Innovative Technologies in Engineering (ICACITE)*, IEEE, pp. 1477–1481, April 2022.
6. Ozbolat, I.T. and Hospodiuk, M., Current advances and future perspectives in extrusion-based bioprinting. *Biomaterials*, 76, 321–343, 2016. https://doi.org/10.1016/j.biomaterials.2015.10.076.
7. Jakus, A.E., Rutz, A.L., Shah, R.N., Advancing the field of 3D biomaterial printing. *Biomed. Mater.*, 13, 3, 034102, 2018. https://doi.org/10.1088/1748-605X/aab058.
8. Skardal, A. and Atala, A., Biomaterials for integration with 3-D bioprinting. *Ann. Biomed. Eng.*, 43, 3, 730–746, 2015. https://doi.org/10.1007/s10439-015-1390-2.
9. Kaswan, K.S., Baliyan, A., Dhatterwal, J.S., Jain, V., Chatterjee, J.M., Intelligent classification of ECG signals using machine learning techniques, in: *Healthcare Monitoring and Data Analysis Using IoT: Technologies and Applications, Decision and Control*, p. 412, 2022.
10. Murphy, S.V. and Atala, A., 3D bioprinting of tissues and organs: A review spanning multiple scales. *Trends Biotechnol.*, 38, 10, 1044–1060, 2020. https://doi.org/10.1016/j.tibtech.2020.03.008.
11. Atala, A. and Yoo, J.J., *Essentials of 3D Biofabrication and Translation*, Elsevier, Edinburgh, London, Oxford, UK, Amsterdam, Netherlands, 2015.
12. Jang, J. and Cho, D.W., 3D bioprinting and modeling for advanced drug delivery. *Chem. Eng. Sci.*, 224, 115831, 2020. https://doi.org/10.1016/j.ces.
13. Li, H., Fan, W., Zhu, X., Three-dimensional printing: The potential technology widely used in medical fields. *J. Biomed. Mater. Res. A*, 108, 11, 2217–2229, Nov. 2020. https://doi.org/10.1002/jbm.a.36979.
14. Kaswan, K.S., Dhatterwal, J.S., Kumar, S., Pandey, A., Industry 4.0 multiagent system-based knowledge representation through blockchain, in: *Artificial Intelligence and Industry 4.0*, pp. 93–115, Academic Press, Cambridge, Massachusetts, 2022.

15. Zhang, Y.S., Aleman, J., Arneri, A., Bersini, S., Piraino, F., Shin, S.R., Khademhosseini, A., From cardiac tissue engineering to heart-on-a-chip: Beating challenges. *Biotechnol. Adv.*, 36, 3, 941–953, 2017.
16. Miri, A.K., Nieto, D., Iglesias, L., Goodarzi Hosseinabadi, H., Maharjan, S., Ruiz-Esparza, G.U., Khademhosseini, A., Fabrication of complex porous tissue constructs using a pressure-assisted bioprinting approach. *Adv. Healthc. Mater.*, 8, 6, 1801011, 2019.
17. Tarakanova, A. and Cherkaoui, S., Bioprinting of vascularized tissues: Principles, promises, and challenges. *Curr. Opin. Biotechnol.*, 55, 90–100, 2019.
18. Kaswan, K.S., Gaur, L., Dhatterwal, J.S., Kumar, R., AI-based natural language processing for the generation of meaningful information electronic health record (EHR) data, in: *Advanced AI Techniques and Applications in Bioinformatics*, pp. 41–86, CRC Press, Boca Raton, Florida, 2021.
19. Zhang, Y.S., Aleman, J., Shin, S.R., Kilic, T., Kim, D., Mousavi Shaegh, S.A., Dokmeci, M.R., Multisensor-integrated organs-on-chips platform for automated and continual *in situ* monitoring of organoid behaviors. *Proc. Natl. Acad. Sci.*, 114, 12, E2293–E2302, 2017.
20. Gu, Q., Tomaskovic-Crook, E., Wallace, G.G., Crook, J.M., Zhang, B., Mechanically patterned neuromuscular junctions-in-a-dish have improved functional maturation. *Integr. Biol.*, 10, 10, 520–531, 2018.
21. Groll, J., Boland, T., Blunk, T., Burdick, J.A., Cho, D.W., Dalton, P.D., Heilshorn, S.C., Biofabrication: Reappraising the definition of an evolving field. *Biofabrication*, 8, 1, 013001, 2016.
22. Dhatterwal, J.S. and Kaswan, K.S., Intelligent agent based case base reasoning systems build knowledge representation in COVID-19 analysis of recovery of infectious patients, in: *Applications of Artificial Intelligence in COVID-19*, pp. 185–209, 2021.
23. Ozbolat, I.T. and Yu, Y., Bioprinting toward organ fabrication: Challenges and future trends. *IEEE Trans. Biomed. Eng.*, 60, 3, 691–699, 2013.
24. Hinton, T.J., Jallerat, Q., Palchesko, R.N., Park, J.H., Grodzicki, M.S., Shue, H.J., Gaudette, G.R., Three-dimensional printing of complex biological structures by freeform reversible embedding of suspended hydrogels. *Sci. Adv.*, 1, 9, e1500758, 2015.
25. Rengier, F., Mehndiratta, A., von Tengg-Kobligk, H., Zechmann, C.M., Unterhinninghofen, R., Kauczor, H.U., 3D printing based on imaging data: Review of medical applications. *Int. J. Comput. Assist. Radiol. Surg.*, 5, 4, 335–341, 2010.
26. Marques, A.C., Soares, C.F., Reis, R.L., Goes, J.C., Biofabrication and computational modeling for tissue engineering applications. *Front. Bioeng. Biotechnol.*, 5, 48, 2017.
27. Sarker, M.D., Naghieh, S., McInnes, A.D., Schreyer, D.J., Chen, X., 3D bioprinting for precision medicine in cancer. *Expert Rev. Precis. Med. Drug Dev.*, 3, 4, 207–220, 2018.

28. Lee, V.K., Kim, D.Y., Ngo, H., Lee, Y., Seo, L., Bioprinting and its future applications, in: *The 3D Printing Handbook*, vol. 2, pp. 27–45, 2020.
29. Chen, C., Bang, S., Cho, Y.S., Lee, S.J., Lee, J.H., Advances in tissue engineering using 3D printing. *J. Mater. Chem. B*, 7, 28, 4433–4455, 2019.
30. Chen, Y., Zheng, X., Yue, K., Liu, L., Chen, X., Zhu, X., Zhang, Z., Bioprinting of complex tissues: From pre-processing to post-processing. *Biofabrication*, 13, 2, 022001, 2020.

# 12

# Computer Vision-Aides 3D Bioprinting in Ophthalmology Recent Trends and Advancements

**Jagjit Singh Dhatterwal[1], Kuldeep Singh Kaswan[2*], Ankita Tiwari[3], Balamurugan Balusamy[4] and R. Gopal[5]**

[1]*Department of Artificial Intelligence & Data Science, Koneru Lakshmaiah Education Foundation, Vaddeswaram, Andhra Pradesh, India*
[2]*School of Computing Science and Engineering, Galgotias University, Greater Noida, India*
[3]*Department of Engineering Mathematics, College of Engineering, Koneru Lakshmaiah Education Foundation, Vaddeswaram, AP, India*
[4]*Shiv Nadar University, Delhi, India*
[5]*Information and Communication Engineering, University of Buraimi, Al Buraimi, Oman*

***Abstract***

In recent years, the field of ophthalmology has witnessed significant advancements in the application of 3D bioprinting technology for the fabrication of ocular tissues and organs. However, the complexity and precision required in ophthalmic bioprinting demand advanced tools and methodologies. This abstract explores the recent trends and advancements in the integration of computer vision techniques to aid 3D bioprinting in the field of ophthalmology. Computer vision techniques, including image processing, pattern recognition, and deep learning algorithms, have shown great potential in assisting various stages of the ophthalmic bioprinting process. These techniques enable the analysis and interpretation of ophthalmic imaging data, such as optical coherence tomography (OCT) and fundus images, to extract critical features for the design and fabrication of ocular structures. One key application of computer vision in ophthalmic bioprinting is the segmentation and reconstruction of ocular structures from medical imaging data. By utilizing image processing algorithms and deep learning models, computer vision techniques can

**Corresponding author*: Kaswankuldeep@gmail.com

E. Gangadevi, M. Lawanya Shri, Rajesh Kumar Dhanaraj and Balamurugan Balusamy (eds.) Computational Intelligence in Bioprinting, (239–258) © 2024 Scrivener Publishing LLC

accurately segment and extract anatomical features, such as the cornea, retina, and optic nerve, from complex ophthalmic images. This information can then be used to guide the 3D bioprinting process, ensuring the precise fabrication of patient-specific ocular tissues. Another area where computer vision techniques prove valuable is in quality control and real-time monitoring of the bioprinting process. By employing computer vision algorithms, researchers can monitor the printing parameters, assess the structural integrity, and detect any deviations or defects during the fabrication of ocular tissues. This enables prompt adjustments and ensures the production of high-quality and functional ocular structures. Moreover, computer vision-based virtual reality (VR) and augmented reality (AR) technologies offer immersive visualization and interaction capabilities, allowing surgeons and researchers to evaluate and manipulate the 3D bioprinted ocular models in a realistic and intuitive manner. This aids in the preoperative planning and assessment of the printed structures, leading to improved surgical outcomes and patient-specific treatments. The integration of computer vision techniques in 3D bioprinting for ophthalmology represents a significant advancement in the field. It enhances the accuracy, efficiency, and reliability of the bioprinting process, facilitating the fabrication of complex and personalized ocular structures. Moreover, the utilization of computer vision-assisted VR and AR technologies provides a more comprehensive understanding of the printed models, aiding in surgical planning and training.

***Keywords:*** Bioprinted, 3D printing, lens and prosthetic, ophthalmology

## 12.1 Introduction

Since its introduction 30 years ago, 3D printing has proven to be an invaluable tool for quick prototyping, production, and tooling. As a result of the many advantages of 3D printers in recent years, their availability to both professional and amateur users has skyrocketed. During the last three years, the worldwide market for printing devices, as well as the materials and equipment used by and their value to other sectors, has expanded by more than 33% [1]. IDTechEx predicted in The Independent that 3D printed cells and tissues will have an estimated capability of $6bn by the year 2025, contributing to a total medical sector growth of $867m. It's remarkable to see how far 3D printers have come to become standard equipment in institutional settings like schools, libraries, and research facilities. Three-dimensional printing works because it is based on the additive manufacturing (AM) theory, which describes the layer-by-layer mixing of material with the assistance of outside stimulating. The both functional and ineffective parts are manufactured using computer-aided design (CAD) and computer-aided manufacturing (CAM) software tools, which leverage manufacturing methods to turn a graphic designing into a physical item [2].

Like 2D printing on flat paper, this method has had widespread uptake. Weapons, apparel, and even high-end accessories like jewelry and vehicle components may all be made using 3D printing. These innovations are very new to the medical industry, yet they might significantly alter the face of surgery and healthcare as we know it. Vascular networks, the production of coronary arteries, dressings, ears, bones, prosthetic limbs, dental prostheses, and eustachian tubes are all areas where culture media and 3D printing have been intensively investigated in the literature because of their prospective uses. The promise of 3D printing will soon be seen in cell biology and materials research. Furthermore, splints, stents, and other clinical medical devices are produced using this technique. Prototypes for all sorts of unique components, types, and designs are created using this technology by the research community, along with structural models with custom designs. The current materials must also conform to particular product criteria, making 3D printing a somewhat expensive way to get customized items [3].

As it can function as a standalone unit, it has found widespread use in the design of complicated components. Moreover, 3D printing has a major effect on the industrial industry by enhancing dependability, build speed, and user-friendly interfaces. Three-dimensional printing has led to the development of a wide range of new materials, with the aim of expanding the reach of manufacturing to sectors such as the scientific community, independent material suppliers, and machine builders. Switching frequency, in additional to the necessity for specialized processing parameters, is the most difficult scenario to overcome while developing a new material. In order to build human-specific structures out of biological or excellent biocompatibility materials, AM methods are now among the most cutting-edge and promising options. Controlling the design of AM's end products is hindered, in part, by the difficulty inherent in precisely positioning a large number of cells, biomaterials, and active chemicals inside narrowly specified outlines. Recent advances in bioprinting technology have allowed for the construction of 2D and 3D structures with extreme accuracy and the precise insertion of cells, biomaterials, and biomolecules at chosen areas, overcoming the aforementioned obstacles [4].

The 3D bioprinter has a maximum temperature of 50°C150°C, allowing it to print a wide range of biomolecules. These materials include ceramics, cells, proteins, hydrogels, biodegradable plastics, etc. Bioprinting was pioneered in 2000 by Thomas Boland of Clemson University, who used a standard ink-jet printer to generate a bio superstructure using a bacteria amalgamation in an ink cartridge. Once his projects were finished and given the go light, he filed for and was granted patents in 2003 and 2006. The Novogen bioprinting technology and the development of cellular bio-ink were both conceived and outlined by Dr. Gabor Forgacs (Professor, Missouri University, USA). It is noteworthy that the development of various bio-ink materials has enabled scientists to precisely modify the *in-vivo* biological and biochemical environments of living things, hence facilitating the cellular manufacture of sophisticated biological inventions.

His team developed and put the NovoGen MMX advanced bioprinter on the market in 2007 under the umbrella company Orgonovo Inc. The high price of this printer greatly reduced its market share. Organ bioprinting has been around since 2003, when it was initially published. From 2003 to 2011, several advancements were achieved in the bioprinting application arena, such as the invention of a three-axis bioprinting system with an injection syringe that is under computer control. Ophthalmology is another potential use of 3D printing technology's broad applicability. Besides printing eyewear, this method may also be used to replicate ocular tissues like the cornea and sclera for scientific study. It has been suggested that 3D printing may be used to create physically accurate bio-models that would greatly improve the study of human anatomy and the diagnosis and treatment of illness. In addition, a customized 3D printed prosthetic eye may be created in a matter of weeks using the latest in 3D printing and 3D scanning technologies. Whole new organs for the treatment of certain disorders may one day be printed using 3D printing technology. Possible uses of 3D printing in ophthalmology include patient-specific prosthetic ocular implants, surgical planning for corneal transplants, and the creation of glaucoma valves [5].

## 12.2 Digital Laser Printing Techniques

There are several 3D printing techniques available, and they may be broken down into the aforementioned people (liquid, solid, powder). Layer beams, solid footing curing, Teijin Seiki's solid form system, Meiko's AM system for the diamond industry, SLP, digital laser-active simulation models machine, layer simulation models system, layer contouring, plastic sheet plastic laminate by Solidimension, paper lamination advanced technologies by KIRA, CAM-LEM'S CL 100, ENNEX corporation's offset, and a plethora of other in-house established technologies have all not maintained their market prices [6].

### 12.2.1 Tissue Engineering Industry

These days, bioprinting technologies are all the rage since they can solve several technical problems plaguing the tissue engineering industry (TE). Critical TE applications involving cells, scaffolds, and other biological molecules may be constructed with the use of these technologies. Using bioprinters provides a simple solution to many of the difficulties and issues associated with the traditional TE method, especially those relating to geometrical complexity and accuracy [7]. These computer-controlled processes may deposit, cure, or crosslink materials in successive layers to produce three-dimensional structures with complex geometries and fine detail. Hydrogels and cells, for instance, may be deposited using a nozzle-based deposition process; printers can cross-link between layers by dropping cell/bio-ink as needed; and synthetic or natural polymers can be cross-linked between layers by selectively irradiating single cells with light. Preprocessing, printing, and postprocessing are the three primary technical phases of the bioprinting process. The first technical phase is called pre-processing, and it involves imaging and computer-aided design to create a blueprint of the tissue or organ design [8]. To print tissues, cells are first seeded onto solid and removable scaffolds, and then bio-molecule growth agents are used to induce tissue production. The printed component is then placed in a bioreactor, a customized chamber, to speed up the tissue formation process. Size, porosity, and node density are only some of the aspects of TE structure that may be precisely manipulated by bioprinters [9].

One of the three primary types of biological printing methods may ensure the accuracy, consistency, and survivability of cell depositing to varied degrees (laser-based, extrusion-based, and inkjet-based). One subgroup of these innovations is the ability to rapidly solidify bio-materials that respond to external stimuli (such as lighting, warmth, or chemical). Almost all bioprinters employ one of three basic approaches: bio-mimicry, example of self, or mini-tissue creation. This study combines the uses, characteristics, workhorse materials, advantages, and disadvantages of extant bioprinters, and outlines their therapeutic potential [10].

#### *12.2.1.1 Printing Biomedical Structure*

Jetting-based bioprinting allows for the printing of 2D and 3D biomedical structures by depositing a series of small droplets of bio-ink onto a substrate in successive layers. The jet-based bioprinters may be categorized depending on the processes used to create the bio-ink droplets (piezoelectric actuator, laser-induced, and compressor). To complete the thermal process, a heat generator raises the temperature within the bio-ink chamber, where a bubble forms and is subsequently released as a small droplet. The piezoelectric actuator pushes the droplet out using a pulse actuator made from a piezo crystal [11]. As heat and piezoelectric delivery methods are the most common occurrences in jetting-based printing technology, commercialized printers solely utilize them. The printed hydrogels as provided on the platform have poorer cell viability, but the laser-induced method can yield relatively high-resolution patterns. Pressure regulator systems use the movement of the opening and closing of a very small valve to generate the droplets. This simplest method allows for the ejection of particles with a volume of 20–100 picoliters with a precision of 20–100 micrometers [12].

Low viscosity chemicals including salt water, CaCl2, thrombin, and fibrogen are used in these bioprinters' bio-inks. Nevertheless, it is challenging to properly create structures that will preserve their form and strength after implantation since jet-printed substances frequently have poor mechanical behavior. The bio-ink, which consists of cells and biomaterials, produces droplets with 10,000 to 30,000 living cells. Scientists modified the inkjet printer so that it could print matrices into which various types of cells could be positioned accurately [13]. This allowed for the construction of complex heterogeneous structures.

Researchers have explored the pros and downsides of printing blood arteries and heart valves using inkjet printers that use hydrogels flexible enough to tolerate physiological conditions.

Despite these limitations, this printer offers a number of positive features, such as its low cost, high resolution, quick deposition rate, and interoperability with a variety of biological materials. While the use of jetting-based 3D printers in ophthalmologists is not new, progress in this area has been modest for a variety of reasons. Several factors contribute to this problem, including a dearth of acceptable workhorse materials, inaccuracies in the printing process, changes in chemical and physical properties as a result of exposure to heat or pneumatic pressure, and, most crucially, the high price of printers. Micrometer-scale, mid-infrared lenses with diameters of 10-350 mm and focal ranges of 10-700 mm were initially constructed for considerable portion by University of Pennsylvania scientists using an inkjet printer half a decade ago. The liquid microstructured droplets were inkjet printed onto a warmed, flat surface, where they cured in place to create the flexible optically lenses. Recent findings from an in-depth study of ARPE19 production and retinal photoreceptors [14] have been made public in an effort to treat a broad variety of retinal illnesses.

The first human eye cells were printed using inkjet technology in 2013, with the goal of developing personalized tissue implantation to cure hereditary and acquired blindness. Due to their vulnerability, all of the rat optic nerve cells used in the pilot study unfortunately perished. It makes sense to utilize the kind of substance that permits simple jetting and precision with little aid from heating or pressure in order to reduce clogging and interruptions in the flow of bio-ink. Nevertheless, developing and researching these substances is a whole new field of study that will take a lot of time and money. However, chemists and materials scientists have a difficulty when bioprinting with jetting since they need to figure out how to optimally deliver structural and functional components in liquid form [15]. Another area that requires further study is the utilization of just liquid forms of biological materials. Alternatively, it may be possible to create a new class of materials by means of a novel deposition technique, including powders, ceramics, and thus mixes. It is conceivable that future versions of jetting manufacturing technology will be used during medical procedures to print retinal diseases components (such as artificial camera lens, ciliary body, sclera, corneas, ocular hypertension valves, and augmentations) in the human body, allowing for a more individualized and *in situ* methodology. Ergonomic enhancements to jetting technique should be available for sale in bioprinters for use in operating rooms [16].

#### *12.2.1.2 Electrochemical Bioprinter*

Extrusion-based bioprinting technologies generate 3D organ structures from a bio-material made of cells floating in hydrogel in a long, uninterrupted thread. These printers are also known as pressure-assisted 3d bioprinting and services and better bioprinters. Tissues such as aortic valves, vascular trees with branches, *in vitro* pharmacological models, and tumor modeling techniques are all within reach thanks to extrusion bioprinters. Extruding biological molecules for bioprinters is often done using pneumatic or electromechanical piston or screwdriver dispense mechanisms [17]. Because to the potential for a lag in flow of materials due to gas compression, piston-driven deposition frequently provides more precise control over bio-ink flow through the nozzle. The amount of cell-loaded hydrogel released may be controlled by adjusting either the pressure applied or the piston movement of a dialysis machine. After 3D printing an organ or structure, it is fixed via physical or chemical techniques. As compared to micro-nozzle bioprinters, extrusion-based printing technology have the advantage of being able to print with a wider range of biomaterials, including thick nanomaterials.

Depositing cells of different sorts was the objective of extrusion-based bioprinting employing a wide range of biodegradable hydrogels. Keratinocytes and adipose-derived pericytes were used in conjunction with gelatin/chitosan hydrogels to fabricate artificial liver tissue utilizing extrusion-type ultimately help. A low self - discharge device with multiple nozzles that can be set to a certain micron size has been developed by the research community (such as precision deposition nozzle, pneumatic microvalve, piezoelectric nozzle, and solenoid valve). Moreover, some scientists have developed multilayered cell hydrogel composites through an extrusion-based method. Cell viability after exposure to compressive stresses during printing was investigated utilizing a rheology-based methodology. Cell viability was lower after extrusion bioprinting compared to inkjet bioprinting (40-86%), and decreased with higher extrusion pressures and nozzles gauge [18].

Due to Hydrogel's incapacity to degrade its encompassing gel matrix, the resulting structures perpetuated poor proliferation and exhibited no evidence of differentiating, diminishing the effectiveness of the medical devices and severely restricting their applications. These printers' broad material compatibility opens up a wide range of potential medical applications. Yet, few individuals have reported using this method in the creation of devices or implantation for retinal diseases treatment. There are likely many reasons why extruded bioprinting technology are not yet being used to fabricate retina and eye wears like prescription lenses, but the reality remains that this is the case [19]. The unclean air generated by the compressor in electrically powered extruded machines accelerates the breakdown of anatomical tissues and must be sanitized. The viscosity of the bio-inks is the second major issue with these printers; systems cannot produce accurate and monitored extrusion with semi-solid or solid bio-ink, making sterilization necessary and easily performed with an autoclave. Ultimately, post-treatments appear necessary due to the reduced surface roughness of the resulting structure. For example, a contact lens printed using an extrusion printer would have a surface that is irritating and unpleasant for the wearer. In this case, the roughness profiles will also affect how much light reaches the retina. While several post-printing technologies have been developed for plastic injection molding printers, the success of bio-printed products is still in question. Keep in mind the malleability of extrusion-based biomaterials with a variety of bio-ink types, their capability for diffusion and reperfusion, and their speed [20] as you work to overcome the aforementioned restrictions and develop this reconfigurable manufacturing tool for ophthalmic applications.

#### 12.2.1.3 Nozzle Free Printing

Tissue engineering are accumulated onto a platform using light beam in laser-assisted biomaterials. For example, the three parts shown in the following image (pulsed laser source, ribbon coated with liquid biological materials that use to deposit over a metal film, and receiving substrate). This technique, which uses UV lasers as sources of energy, might be used to print hydrogels, cells, proteins, and ceramics. The settlement at which epithelial tissue can print normally varies from pico- to micro-scale functionalities, and the final product's print is impacted by many input information such as the thickness of the bioactive molecules on the film, thermodynamic properties, energy of the laser, and permeability of substrate. This technique has been used to print a wide variety of cell types, including human fibroblasts from the dermis, mouse C2C12 myoblasts, bovine pulmonary artery endothelial cells (BPAECs), breast cancer (MCF-7) cells, and rat neural stem cells. The production of Graftskin skin substitutes using this technique in 2013 was a major step forward for laser-assisted bioprinting [21].

After constructing skin structures from hepatocytes and cellular membranes, we inserted them into the folds of mouse skin. Eleven days after the skin incision, the graft had successfully adhered to the surrounding tissue. The capacity to print extremely engaged cells at full definition and fine control over the location of ink droplets set it apart from other bioprinting technologies. The printing process itself is nozzle-free and non-contact. Nozzle-free printing allows for the use of ensure the development, something not yet achievable with conventional bioprinters. Neuroscientists have shown that mammalian cells can be printed onto a hydrogel platform in either stack or as single cells, and lasers make it easy to print even the minute elements of an organ. The usage of laser printers to implant cells and form cellular stacks with a thickness of 50-100 lm was discovered by the research team. Nanohydroxyapatite was bioprinted *in vivo* using a mouse calvaria 3D defect model with a 3 mm diameter and 600 mm deep hole. In a case of localized internal and external internal [22], a medical device (such as a customized noncellular bioresorbable tracheal splint) created with a laser was implanted into a young child.

There is only so much that can be done with current technology until it advances. The heat generated by laser light has the potential to harm cells and their ability to interact and form aggregates. Laser-based bioprinting has the additional problem that cell placement in the precursor solution is determined by gravity and chance. Photo cross linkable biomaterials are essential; however, the production process is lengthy, and 3D printing is currently not an option. Laser-based bioprinting techniques have proven to be the most successful in the area of endocrinology. Laser-assisted painting is the most expensive method, and it may be time-consuming and incorrect when putting cells, although allowing for the printing of concentrated fluids without compromising cell viability. This method, unlike earlier ones, does not need the use of a nozzle, making it suitable for the use of even high-viscosity inks [23].

As a result of the method's precision manufacturing capabilities, it can compete well with other technology that show promise for use in ophthalmology. Laser-based systems biology has a variety of drawbacks that researchers are constantly working to address. Yet, there is little to no indication of active research into how to overcome the challenges of developing multi-cell type integrating and making sure the laser prints last for a long time. If this technique is to serve patients, the exorbitant cost of laser printers must be reduced. This is impossible without a transfer of technical rights from 3D printer manufacturers to end users. If this is accomplished, it will pave the way for restoring sight for the purposes of visual rehabilitation and increasing the standard of living for the blind.

## 12.3 3D Printing Biological Material

Bioprinting techniques, which employ 3D printing to manufacture biological material, have the potential to revolutionize the field of ophthalmology. More than 30% of people throughout the world have some kind of visual impairment due to conditions including refractive errors and cataract. The eye may be operated on and implants implanted with relative ease despite being a complex organ. As a result, bioprinting might be used to meet the complex requirements of ophthalmic treatment. Many considerations need to be made before bioprinting techniques may be used in ophthalmology. The 3D-printed structures need to be biodegradable to promote cell attachment, mechanically robust to withstand the mechanical load within the eye, and sized appropriately for use there. Bioprinting may be done without using live cells. Cells are often used after being encased in a hydrogel for use. A recent literature study found that just five clinical trials have investigated the use of bioprinting technology in ophthalmology. Even in their infancy, these innovative technologies have already paved the path for some significant breakthroughs in ophthalmologist. Bioprinted lenses and prosthetics, cell printing, ocular disorders, drug delivery, and technological advances like the bionic eye are all covered in this article [24].

### 12.3.1 Optical Quality of 3D Printing Technology

Due to the high level of manufacturing accuracy and extremely precise finishing requirements, producing individualized optical devices is prohibitively costly and unfeasible. High-resolution transparent polymers with comparable optical qualities have recently become possible because to advancements in 3D printing technology. Optical elements may now be designed and fabricated quickly using 3D printing technologies for far less money than traditional manufacturing, expanding their availability and decreasing their end-to-end prototype time. It has also been stated that AM has been used to create flexible optical lenses for usage in cellphones. The low-cost photopolymer printing technique that underpins printed optics has been proven, and it is already being employed by a variety of entry-level 3D printers.

Precision AM technologies reportedly make it feasible to create sophisticated models of a patient's eye anatomy, which will improve the accuracy and success rate of surgical interventions. There just isn't enough evidence from studies in this area to back up these assertions. In addition, a replacement for the 3D empty eye model that was made using an AM machine about 10 years ago to evaluate innovative healing difficulties in therapies for retinal illnesses has not yet been identified. There aren't a lot of usable AM materials for ophthalmology now, therefore there's a lot of room for innovation in the field of ocular tissue research (such as conjunctiva, sclera and corneas). Making custom eyewear for people with facial deformities using 3D printing technology is a step toward creating customized prosthetics like artificial lenses and glaucoma valves. The prospective applications of bioprinting in eye care are really intriguing. In ophthalmology, 3D printing has been the subject of many patents [25].

Most of the patents deal with the use of 3D bioprinting to ophthalmology, namely the production of ophthalmic lenses. Engineers at the United Kingdom's Manchester City University have developed and patented a method for fabricating artificial eyeballs that may be implanted into patients to restore their eyesight in whole or in part. This patent outlines a multistep procedure for fabricating artificial eyes: To create an artificial eye, first a 3D picture of an iris is printed onto a substrate, then the front of the mechanical eye is supported, the substance and supporter are placed within a mold, and last the mold and supporter are enclosed in a mound. Customized and on-demand printing of surgical devices such prosthetic lenses, glaucoma valves, and other prosthesis is seen as a realistic possibility in the near future. Moreover, this method may be used to enhance a variety of advanced ophthalmological tools in the future.

### 12.3.2 Repair Damage Tissue

TE is an attempt to solve a wide range of issues associated with deterioration and malfunction as a result of both natural and artificial causes. In this branch of engineering, surgeons are increasingly relying on a combination of new technology and biomaterials to repair or replace damaged tissues or organs via the calculated placement of healthy cells with normal functional potential. Literature reports abound on the topic of using 3D printing to create complex cellular structures by using methods like laser-assisted printing, inkjet printing, and microextrusion printing to create functional corneal substitutes [26].

The research of illness recovery and the exploration of artificial network properties, both of which rely on brain cells, are driving the development of artificial dopaminergic neurons. Because to recent advancements in cell patterning, it is now much easier to place brain cells within a microsized chamber or to produce a certain form (such as: soft-lithography, ink-jet printer, bio-plotters, etc.). The combination of 3D printing and a radial electrospinning scaffold for axon guidance control may improve ganglion cell implantation via cell-specific orientation. Researchers at the University of Cambridge fed components of rat retinas into a conventional ink-jet printer to produce multilayer constructions made up of two kinds of cells. Preliminary results suggested that the ink-jet printing process had no negative impact on cell viability or their ability to multiply in culture. In a first for 3D printing, researchers have successfully created a fully functional nervous system from adult animal cells. Previous work has shown that embryonic neural cells may be successfully ink-jet printed; future work will extend this to printing light-sensitive synapses and photoreceptors epithelial cells. Ocular surgery to restore injured retinas may one day be possible thanks to the expansion of this method, which has researchers enthusiastic.

Several buffers, ejection energies, and cell densities were tested to see how thermal ink-jet printing affected the growth and survivability of retinal ganglion cells *in vitro*. Thermal print heads should be used with care, despite the fact that several investigations demonstrate that the heating of these print heads has no impact on the continued existence of various cell types, including cell lines, hamsters ovary cells, muscle, and stem cells. Cells from the adult rat central nervous system, retinal mesencephalon, and glia have all been studied for their survivability and potential for development when printed using ink-jet printers. It has been shown that ganglion cells and glia may be printed successfully using a piezo-electric printer. Yet, the piezoelectric process caused a reduction in flow cytometry in areas with higher shear stresses. Progress in creating a functioning retina is being used to the development of a sophisticated 3D cellular architecture [27].

The transmission of image perception from the eye to the brain also requires cooperation between retinal cells. When certain cell types need to be replenished, the danger of primary cell loss arises (as in glaucoma or pigmentosa). Surgical implantation of a 3D-printed macula might replace the need to remove and replace a whole eyeball in cases with advanced retinoblastoma. One possible solution to the latter issue is creating a 3D duplicate of the human eye that is devoid of cells. Here is where ink-jet printing shows promise as a potential treatment for degeneration blindness by producing new tissue.

### 12.3.3 Eye Blindness

Medical screen printing and 3D bio-printing methods may have significant applications in ophthalmology. Some of the most common causes of blindness include retinal vein occlusion (RVO), myopic choroidal neovascularization (mCNV), and age-related macular degeneration (AMD). Age-related macular degeneration (AMD) is becoming more common as lifespans increase all around the globe.

If left untreated, neovascularization (a hallmark of these retinal diseases) caused by vascular endothelial development factors rapidly progresses to total blindness (VEGF-A). In response to hypoxia and other stimuli, vascular endothelial growth factor-A (VEGF-A) binds to VEGF-A receptors and plays a critical role in the formation of both healthy and unhealthy blood vessels inside the eye (such as Flt-1). Because of their capacity to block the VEGF activity that leads to CNV, anti-VEGF antibodies have replaced previous treatments for AMD, DR, mCNV, and RVO. This procedure may enhance eyesight for some individuals while leaving others with permanent harm such CNV leaking, scarring, or geographical atrophy. Because the eyesight of one-third of those who have underwent eye surgery does not increase much, they are considered legally blind. Hence, a novel therapeutic approach that improves results while retaining a satisfactory long-term safety is urgently needed. Bioprinting technology might be used to recreate photoreceptors epithelial, a kind of ocular cell that is lost in retinal diseases including AMD and DR. The retina is made up of many layers of sensory neurons, each of which plays an essential role in the retina's overall function. Retinal injury healing requires meticulous placement of cells in a well-planned 3D architecture. Primary open-angle conjunctivitis is the leading cause of blindness and visual impairment, affecting more than 70 million people throughout the globe (POAG). Without treatment, it will ultimately lead to blindness due to the loss of retinal ganglion cells (RGCs). The most common form of glaucoma, known as primary open-angle glaucoma (POAG), is linked to advanced age and high intraocular pressure (IOP).

Yet, despite initially responding well to medication, a sizable minority of persons with POAG continue to lose vision because they are resistant to those medicines and/or don't use the eye drops as suggested, resulting to continuous chronic disease. Thus, it is crucial to actively look for and develop novel methods of RGC replacements. The manufacturing of two types of neural cell using retinal cells has been declared as an accomplishment by researchers at the University of Cambridge in the United Kingdom. Glial cells and ganglion cells work together to insulate, protect, maintain, and nourish the neurons that make the central nervous system and transmit visual information to the brain. Each team used piezoelectric inkjet-printed gel to determine how well their cultures were propagated. The piezoelectric printer is seldom used for bio-printing since it uses electric currents to produce ink droplets, a process that might potentially harm cellular membranes. Sedimentation inside the printer head reduces the number of cells, but does not influence their vitality. Moreover, printed cells showed similar regeneration properties and overall survival in the composing medium as their nonprinted counterparts. Monolayers of printed glial cells and printed retinal cells were put on top of printed ganglion cells and photoreceptors. For a functional retina to be built, a complex three-dimensional cellular structure that can exchange cell signals must be generated.

Two smartphone-compatible gadgets that use 3D printing in ophthalmologist have recently been released. This 3D-printable imaging adapter allows you to take pictures of your retina using the camera on your smartphone. One such superior image of a glaucomatous disc captured by the device over a standard fundus camera. Using the same design ethos, the group also made a slit light microscope adapter. This device may be attached to a cellphone camera to take pictures of the front of the eye, where conditions including cataract, uveitis, ocular ulcers, and epithelial cell irregularities can be clearly seen. These devices have a price tag that is less than a tenth of that of standard ocular imaging equipment.

According to the American Transplant Organization, corneal transplantation are done on over 40,000 Americans each year, making them the most prevalent kind of organ transplantation in the country. Around 53% of the world's people are without access to corneal organ transplants, in accordance with global research published in *JAMA Ophthalmology* in February 2016. The transplant that had been inserted was rejected because of issues with the innate immunity. The anatomy of the eye and orbit is quite complex. Anatomical models generated on a 3D printer and used for education and training, as well as to clarify the anatomical relationships between normal tissues and lesions, are quite helpful in these circumstances. This cutting-edge innovation has considerably aided both formal education and medical practice. Using a 3D Systems Z650 printer, scientists are creating "extremely accurate" 3D replicas of the orbit, which improves vision of the sensitive optic nerves.

### 12.3.4 Medicine Company

With the advent of cutting-edge technologies like 3D printing in the area of ophthalmologist, the possibility of connecting the nervous system of a person to electrical components like mechanical parts in the quest to cure blindness is moving closer to reality. Because of advancements in this field, high-tech bionic eyes are now within reach. One of the most promising areas of research in pharmaceutical companies in recent decades has been the use of nanoparticles to increase ocular residence time and therapy while enhancing ocular bioavailability and lowering drug toxicity. Many nanotechnology-based carrier systems, including nanosuspensions, nanoparticles, liposomes, and Nano micelles, have been developed and studied in depth. Drug release is being sustained, targeting can be improved (where targeting moieties are employed), and fewer doses are needed thanks to nanocarriers and nanodevices. Non-invasive medication delivery to particular parts of the eye, including the back of the eye's tissues, still requires the development of a carrier system. It's evident that research into better ways to administer medications is warranted [28].

While these polymeric systems with branching structures have shown promise in other contexts, their potential for ocular medicine administration has not yet been adequately investigated. Drug and protein-loaded biomaterials may be crucial in the fight against eye disease. Particularly, a 3D-printed drug delivery device is customized with a patient's specific medication mix. NTU (Singapore) has welcomed the use of lipid nanoparticles as a drug delivery technique for the production of extended release preparations of ocular drugs and for transporting them to various anatomical areas of the eye. Liposomes have been used effectively to transport prostaglandins, antibiotics, and progovernment drugs to the front and back of the eye for long periods of time.

### 12.3.5 Artificial Prosthetic Eye

To restore functional vision to people who have lost it due to partial or complete blindness, scientists have created the bionic eye, also known as a visual prosthesis, which consists of an artificial visual system that includes associated equipment. Some 124 million individuals around the world have low vision, as well as another 40 million are totally blind, according to the latest estimates. There are several significant differences between a bionic eye and a prosthetic eye. Prosthetic eyes are used to provide those who have lost an eye to disease, trauma, or birth defects a chance to regain their normal appearance and function. Bionic eyes, which are on the other hand, are prosthetic implants that are surgically inserted into a person's brain or eye sockets. There have been significant breakthroughs in numerous bionic eye implants, however at present only one prosthetic limb system can treat blindness difficulties largely caused by certain eye illnesses and is commercially accessible in the United States. Using evok3d and three dimensional Systems' ProJet 1200 3D printers, scientists at the Bionics Institute have developed a method of 3D printing that could restore vision. Coupling the bionic eye to the visual cortex, optic nerve, and retinal ganglion cells has been recognized as a significant barrier to its clinical use. These concerns need more research. By 2020, it may be possible to introduce the bionic eye to the general population. The Italian company MHOX is working on a bionic eye with a base that is inserted in the eye socket and communicates with the visual cortex and the brain via 3d printing technology tissue.

## 12.4 Conclusion and Future Work

In comparison to other biomedical applications, research into the use of bioprinting technology in ophthalmology is still in its infancy. Some examples of the high-tech and delicate ophthalmology equipment developed by ophthalmologists include corrective lenses, artificial eyes, eyeglasses, retina conjunctiva, sclera, corneas, hypertension valves, and efficient digital enabled eye wear. Yet, no study has yet revealed a certain method to help bioprinting technology. This may be because to the lack of bioprinters capable of producing high-quality ophthalmic gadgets. Commercial biomaterials systems have a number of challenges, such as a deficiency of suitable biomaterials, an unsuitable represents a clear, mechanical constraints, a sluggish printing speed, and a high price tag. The development of a fully functional prosthetic eye, made possible by bioprinting-assisted ophthalmologist, is becoming closer all the time. In addition, bioprinting technology should be transferred from the laboratory to the bedside, where they may be used to develop *in-situ* strategies that mimic the normal eye's structural components while also improving their performance. It is important to choose the best printing parameters and appropriate materials to guarantee the printed devices will last. The ophthalmic industry and its teaching faculty should work to improve the accuracy and viability of nanostructure manufacture in order to expand the use of bioprinting technology in the production of nanostructures like the retinal and contact lens. Finally, the technology rights should be split between the creators and commercial retinal diseases manufacturers so that the equipment and items may be made available to visually impaired people at reasonable cost. There have been some exciting new discoveries in the design of cellular 3D structures, and they may have contributed to the success of 3D printed retinas. Printing cells with the required morphology but without compromising cell viability is possible using glia, retinal ganglion cells from adult rats, and retinal cells from numerous species. Further research is needed to apply these findings to human tissue and other types of retinal cells. There are a number of issues that need to be resolved before bioprinting systems may flourish, despite their broad application in the biomedical and TE fields. Problems with spatial organization, cell density, long-term cell survival, and functional integration of multiple cell types are only a few of the design problems presented by complex 3D printed tissues. In addition, this method may be used with the use of advanced electrospun surfaces in the creation of future retinal models or treatments. Research potential in building an implantable 3D printed retina requires engineering creativity, regulatory effort, and financial commitment. Using x-rays of the eye, doctors are able to construct implants with the exact same dimensions as the patient's native eye. Additive printing allows for the exact patterning of a patient's own cells and appropriate materials on a specified substrate, allowing for the creation of implants that are unique to each patient while reducing the likelihood of rejection. It is also acknowledged that 3D bio-printing may be used to accommodate the large diversity of human body shapes. Research into the use of 3D printing in ophthalmology has led researchers to believe that this technology might help surgeons treat visual impairment patients more safely. The primary goal is to get insight into the rapidly growing field of medical technology customization. 3D-printed ophthalmological devices may one day help those who have trouble seeing or have other eye problems.

## References

1. Fan, D., Li, Y., Wang, X., Zhu, T., Wang, Q., Cai, H., Li, W., Tian, Y., Liu, Z., Progressive 3D printing technology and its application in medical materials. *Front. Pharmacol.*, 11, 122, 2020. https://doi.org/10.3389/fphar.2020.00122.

2. Ali, M., Pages, E., Ducom, A., Fontaine, A., Guillemot, F., Controlling laser-induced jet formation for bioprinting mesenchymal stem cells with high viability and high resolution. *Biofabrication*, 6, 4, 045001, 2014.
3. Zein, N.N., Hanouneh, I.A., Bishop, P.D., Samaan, M., Eghtesad, B., Quintini, C., Miller, C., Yerian, L., Klatte, R., Three-dimensional print of a liver for preoperative planning in living donor liver transplantation. *Liver Transplant.*, 19, 12, 1304–1310, 2013.
4. ASTM F2792-12a, *Standard Terminology for Additive Manufacturing Technologies, F42.19*, ASTM International, West Conshohocken, PA, 2012.
5. Ashammakhi, N., Ahadian, S., Xu, C., Montazerian, H., Ko, H., Nasiri, R., Barros, N., Khademhosseini, A., Bioinks and bioprinting technologies to make heterogeneous and biomimetic tissue constructs. *Mater. Today Bio*, 1, 100008, 2019.
6. Bakhshinejad, A. and D'souza, R.M., A brief comparison between available bio-printing methods. *Biomedical Conference (GLBC)*, IEEE Great Lakes, 2015, IEEE, pp. 1–3, 2015.
7. Barron, J.A., Wu, P., Ladouceur, H.D., Ringeisen, B.R., Biological laser printing: A novel technique for creating heterogeneous 3-dimensional cell patterns. *Biomed. Microdevices*, 6, 2, 139–147, 2004.
8. Mohamdeen, Y.M.G., Tabriz, A.G., Tighsazzadeh, M., Nandi, U., Khalaj, R., Andreadis, I., Boateng, J.S., Douroumis, D., Development of 3D printed drug-eluting contact lenses. *J. Pharm. Pharmacol.*, Dec. 20, 2021. https://doi.org/10.1093/jpp/rgab173.
9. Adams, W., Paxton, L., Dawes, K. *et al.*, 3D printed reproductions of orbital dissections: A novel mode of visualising anatomy for trainees in ophthalmology or optometry. *Br. J. Ophthalmol.*, 99, 9, 1162–1167, 2015.
10. Su, A. and Al'Aref, S.J., History of 3D printing, in: *3D Printing Applications in Cardiovascular Medicine*, ed 1, S.J. Al'Aref (Ed.), pp. 1–10, Academic Press, Boston, 2018.
11. Billiet, T., Vandenhaute, M., Schelfhout, J., Van Vlierberghe, S., Dubruel, P., A review of trends and limitations in hydrogel-rapid prototyping for tissue engineering. *Biomaterials*, 33, 26, 6020–6041, 2012.
12. Alam, M.S., Sugavaneswaran, M., Arumaikkannu, G., Mukherjee, B., An innovative method of ocular prosthesis fabrication by bio-CAD and rapid 3-D printing technology: A pilot study. *Orbit*, 36, 4, 223–227, 2017.
13. Anderson, J., Wealleans, J., Ray, J., Endodontic applications of 3D printing. *Int. Endod. J.*, 51, 9, 1005–1018, 2018.
14. Bradshaw, S., Bowyer, A., Haufe, P., The intellectual property implications of low-cost 3D printing. *Scripted*, 7, 1, 5–31, 2010.
15. Castrejon-Pita, J.R., Baxter, W.R.S., Morgan, J., Temple, S., Martin, G.D., Hutchings, I.M., Future, opportunities and challenges of inkjet technologies. *At. Sprays*, 23, 6, 541–565, 2013.
16. Baliyan, A., Dhatterwal, J.S., Kaswan, K.S., Jain, V., Role of AI and IoT techniques in autonomous transport vehicles, in: *AI Enabled IoT for Electrification*

*and Connected Transportation*, pp. 1–23, Springer Nature Singapore, Singapore, 2022.
17. Dhatterwal, J.S. and Kaswan, K.S., Securing big data using big data mining, in: *Data Driven Decision Making using Analytics*, pp. 1–14, CRC Press, Boca Raton, 2021.
18. Jung, J.W., Lee, J.S., Cho, D.W., Computer-aided multiple-head 3D printing system for printing of heterogeneous organ/tissue constructs. *Sci. Rep.*, 6, 1–9, 2016.
19. Chien, K.B., Aguado, B.A., Bryce, P.J., Shah, R.N., *In vivo* acute and humoral response to three-dimensional porous soy protein scaffolds. *Acta Biomater.*, 9, 11, 8983–8990, 2013.
20. Wu, M., Phoha, V.V., Moon, Y.B., Belman, A.K., Detecting malicious defects in 3D printing process using machine learning and image classification. *ASME International Mechanical Engineering Congress and Exposition Proceedings*, vol. 14, pp. 4–9, 2016.
21. Dhatterwal, J.S., Kaswan, K.S., Pandey, A., Implementation and deployment of 5G-drone setups, in: *The Internet of Drones*, pp. 47–64, Apple Academic Press, Boca Raton, USA, 2022.
22. Dhatterwal, J.S., The role of MAS based CBRS using DM techniques for the supplier selection. *International Journal of Computer Sciences and Engineering*, 7, 5, 2347–2693, Boca Raton, USA, 2019.
23. Yu, C. and Jiang, J., A perspective on using machine learning in 3D bioprinting. *Int. J. Bioprinting*, 6, 4–11, 2020.
24. Zhao, H., Xu, J., Zhang, E., Qi, R., Huang, Y., Lv, F., Liu, L., Gu, Q., Wang, S., 3D bioprinting of polythiophene materials for promoting stem cell proliferation in a nutritionally deficient environment. *ACS Appl. Mater. Interfaces*, 13, 25759–25770, 2021.
25. An, J., Chua, C.K., Mironov, V., Application of machine learning in 3D bioprinting: Focus on development of big data and digital twin. *Int. J. Bioprinting*, 7, 342, 2021.
26. Fahad, M., Dickens, P., Gilbert, M., Novel polymeric support materials for jetting based additive manufacturing processes. *Rapid Prototyp. J.*, 19, 4, 230–239, 2013.
27. Kaswan, K.S., Dhatterwal, J.S., Kumar, S., Lal, S., Cybersecurity law-based insurance market, in: *Big Data: A Game Changer for Insurance Industry*, pp. 303–321, Emerald Publishing Limited, Bingley, West Yorkshire, England, 2022.
28. Fedorovich, N.E., Alblas, J., Hennink, W.E., Öner, F.C., Dhert, W.J., Organ printing: The future of bone regeneration? *Trends Biotechnol.*, 29, 12, 601–606, Bingley, West Yorkshire, England, 2011.

# 13

# Intelligent Image Classification for 3D Printing in Industry 4.0

**Rajbala[1], Kuldeep Singh Kaswan[2*], Jagjit Singh Dhatterwal[3], Gangadevi E.[4] and Balamurugan Balusamy[5]**

*[1]Department of Mechanical Engineering, Galgotias University, Greater Noida, India*
*[2]School of Computing Science and Engineering, Galgotias University, Greater Noida, India*
*[3]Department of Artificial Intelligence & Data Science, Koneru Lakshmaiah Education Foundation, Vaddeswaram, Andhra Pradesh, India*
*[4]Department of Computer Science, Loyola College, Chennai, India*
*[5]Shiv Nadar University, Delhi, India*

## *Abstract*

The rapid growth of 3D printing technology has revolutionized manufacturing processes, enabling the production of complex and customized objects with reduced time and cost. In the context of Industry 4.0, the integration of computational intelligence techniques in image classification for 3D printing has gained significant attention. This abstract explores the potential applications and benefits of using computational intelligence for image classification in the context of 3D printing within Industry 4.0. Image classification plays a crucial role in 3D printing, as it involves converting two-dimensional images or designs into printable three-dimensional objects. However, traditional image classification methods may struggle to accurately interpret complex designs and distinguish intricate details. This is where computational intelligence techniques, such as machine learning, neural networks, and evolutionary algorithms, can be leveraged to enhance the accuracy and efficiency of image classification for 3D printing. By employing machine learning algorithms, computational intelligence models can learn from vast datasets, enabling them to classify images with higher precision and reliability. Neural networks, in particular, offer powerful tools for image recognition and classification, allowing for the identification of intricate patterns and features necessary for successful 3D printing. Furthermore, evolutionary algorithms can be

**Corresponding author*: kaswanrajbala86@gmail.com

E. Gangadevi, M. Lawanya Shri, Rajesh Kumar Dhanaraj and Balamurugan Balusamy (eds.) Computational Intelligence in Bioprinting, (259–274) © 2024 Scrivener Publishing LLC

utilized to optimize the design and fabrication processes in 3D printing. These algorithms can explore a range of design parameters and identify the most efficient configurations, leading to improved print quality, reduced material waste, and enhanced overall productivity. In the context of Industry 4.0, the integration of computational intelligence-based image classification for 3D printing offers several advantages. It enables automated and intelligent decision-making processes, reducing human intervention and potential errors. Additionally, the use of these techniques contributes to the advancement of smart manufacturing systems by facilitating real-time monitoring and adaptive control of the printing process. However, challenges such as data quality, scalability, and interpretability need to be addressed when implementing computational intelligence-based image classification in 3D printing. Ensuring the availability of high-quality training data, scalability for large-scale manufacturing, and interpretability of the classification results are crucial factors for successful integration.

***Keywords:*** Computational intelligence, 3D printing, 3D scanning, artificial neural networks, Industry 4.0

## 13.1 Introduction

Additive manufacturing (three-dimensional printing, 3D printing) is crucial in an Industry 4.0 environment because it provides a revolutionary method of industrial production that opens the way for the creation of lighter, stronger components and systems. Additive manufacturing includes placing new material onto an already existing thing, as opposed to the subtractive methods used in traditional item creation such as milling, machining, carving, and sculpting. The rise in popularity of 3D printing in recent years [1] may be attributed to the use of computer-aided design (CAD) software or 3D object scanners to guide the apparatus in installing material layer by layer in perfect geometric shapes.

During the 1980s, tremendous technical development has been possible thanks to the transition from analogue to digital systems. The introduction of 3D printing, which provided digital adaptability and productivity to the manufacturing process, followed the digitalization in communications, photography, architecture, and engineering. As a result, advanced skill is required to manufacture products utilizing additive methods such as 3D printing. Computing intelligence is required to resolve the aforementioned issues in order to further improve industrial efficiency (CI). Important rules in 3D printing operations may be adopted/optimized using CI. The amount of work that must be done by hand will decrease as faster and better algorithms replace them. Access to big data (massive amounts of information) will be made possible by CI, allowing for more efficient management of 3D technologies such as scanning, printing, and reversing. Thus, CI is an obvious answer to computer-related issues [2].

## 13.2 Advantages

Advantages are better performance, complicated geometries, simpler manufacturing, quick prototyping, easy replication. There are a number of drawbacks, including a lengthy learning curve, a wide variety of skills, a slow individual output rate, and climate change.

Prior to their implementation in CI settings, 3D printing has to have its restrictions addressed. This is especially true when improving the first stages of printing by combining knowledge from other disciplines, such as materials science, 3D printer technology, process design, and programming languages. Problems can be solved more creatively if one has a firm grasp on material properties, as well as how they behave during printing and within the item, on the best materials and 3D printing methodologies to use for a runtime environment, on the design rules for a particular printed constituent, and on the enhancement of specifications. Visual appearance, surface texture and color, repeatability, manufacturing time, and prices [3] are just some of the many factors that must be taken into account in addition to geometrical correctness and mechanical qualities. If the suggested material meets all needs but takes too long to provide, for example, it may be an impediment to the lean management model, and this may be due to circumstances that are difficult to forecast. There are several technical concerns to think about when it comes to 3D printing, and they all have to do with the specifics of the application area. To begin, there are a plethora of technological options:

- Direct Metal Laser Sintering (DMLS)
- Direct Metal Laser Melting (DMLM)
- Electron Beam Melting (EBM)
- Sintering
- Stereo Litho Graphy (SLA)

There are several different ways that materials may be processed, including extrusion, jetting, sheet lamination, vat polymerization, powder bed fusion, binder jetting, directed energy deposition, vat polymerization, and sheet lamination.

Thermoplastics include acrylonitrile butadiene styrene (ABS), polyvinyl chloride (PVC), polyvinyl alcohol (PVA), polycarbonate (PC), polylactic acid (PLA/PLA+), and polyvinyl alcohol (PVA).

Materials including zirconia, alumina, tricalcium phosphate, and powdered glass are used in ceramics, whereas bio-inks, silicon, calcium phosphate, and zinc are employed in biochemicals. • Precious and base metals (gold, silver, steel, titanium, aluminum).

The purpose of this article is to explore whether or not the current potential for using CI to develop 3D printing and reverse engineering has been completely achieved, and if not, where the untapped resources may be found.

## 13.3 Methodology

By surveying previous works, this study investigates CI-based methods that may improve the quality of 3D printing choices.

While training an ANN, we utilized MATLAB 16.0. Input variables were normalized by taking their ranges from the sample data's maximum and lowest points. Initial network weights were predicted to be uniformly distributed between −1 and 1. They were harmonized such that the initial weight doesn't arbitrarily deviate from the startup weight. When deciding when to end learning, both 1000 and 5000 iterations were taken into account. An eye was kept on the mean squared error goal. Each time the least network load was sought, experiments were conducted to determine the optimal combination of inputs, outputs, and hidden layer neuron counts. Statistical software was used to examine the information included in an Excel spreadsheet [4].

## 13.4 3D Printing Technology

- The literature study suggests that CI may be useful because it enables machines to solve problems without human involvement by drawing on their own knowledge and experience. Using this with 3D printing technologies should boost 3D printer performance by lowering mistake rates and making automated manufacturing easier, so it's definitely something to keep an eye on. A key consideration is that we do not have a lot of time to steadily develop and explore. We need to work efficiently and effectively.

Key areas of assistance that rely heavily on CI include:

- Material development,
- Optimization in Parametric,

- Optimization in energy field,
- Less errors as a result of thorough technical examination during the entire manufacturing operation,
- It is feasible to computerize several steps of the 3D printing process, including quality prediction, material and air impurity alarming, and material and air impurity elimination. The process starts with making a CAD file of the design and ends with printing and preparing the model for printing in a slicing application.

### 13.4.1 Automating Selection of Best Tasks

Methods for 3D printing that are intelligent, convenient, slightly elevated, displacement, and service-oriented are being created with the use of intelligent machines at an exponential speed [5].

Data collecting, construction planning, and cost monitoring are just some of the manual processes that may be automated using CI [6]. By increasing machine utilization and rescheduling production orders in light of available resources, CI may help maximize production capacity [7]. Although process improvement has many advantages, it also has certain negatives that must be taken into account, such as inefficiencies and supply constraints in the event of a pandemic. CI can also automate material selection by suggesting the optimal material to use based on the needs of the printed item [8]. We can cut the typical 30-minute time required to prepare a 3D print down to as little as 5-30 seconds, depending on the nature of the job. If we can cut down on this period of time, we can boost printer utilization, decrease mistake rates, and shorten feedback loops [8]. Using CIs simplifies the labor-intensive procedure while improving output quality. In the past, model molds were carved from wax, cast in epoxy resin, or machined from wood. This process might take many days of work, depending on the degree of complexity of the commodity. Digitization of this process has led to significant developments in recent years [10]. These days, master forms are created using CNC milling machines from polyurethane blocks that were originally drawn using computer-aided design (CAD) software.

Smart, economical, high-quality, mass-production, service-oriented 3D printing procedures are studied, planned, optimized, and controlled with substantial use of AI.

Materials and printing process parameters should be selected with lifetime, electricity consumption, and environmental consequences (waste generation and air pollution) in mind [11]. This involves ensuring sure the 3D model can be produced before commencing the print process.

Using parallel slicing algorithms to boost the productivity of prefabrication slicing; Path enhancement, demand correlating, and distribution of resources for on-demand 3D printing services for clients; Remote access to a shared pool of resources (including cloud technology); Detecting and fixing defects at every stage of production. Completing the picture, product lifecycle analysis that takes recycling into account increases security by monitoring production and hardening it against cyberattacks [12], as well as forecasting future risks to the product and the equipment operating the manufacturing line.

The tactic enables artificial intelligence–driven optimization of 3D printing within the framework of Industry 4.0, which in turn lowers the bar for product design and ambiguity, quickens building information modeling, enables real-time control, boosts safety, and identifies flaws at every manufacturing stage [9].

### 13.4.2 3D Printing Performances Analysis

Tuning all process parameters is necessary for the processing of novel, high-performance materials. Here, many sensors are used to keep an eye on the 3D printing process. With the use of CI, we analyze the flow of data and unearth previously unknown connections between variables. Because of this, CI is useful for processing massive volumes of data fast, which is otherwise impossible for humans to do. This allows scientists to foresee the properties of materials and create complex alloys. It is possible to assess the building's print quality before commencing an operation (and incurring production costs). We may also manage the method for minimizing printing errors and waste by making educated guesses about the cleanliness of the parts. This will allow 3D printing departments to tightly regulate their manufacturing processes and ensure the quality of their output. As the market shifts towards real-time, completed components, this is also becoming more vital. Color capture makes it simpler to do things like develop lifelike images during the design phase, and CI can evaluate both the geometry and color data of an item by following its shape and texture with the use of a scanner.

To solve the issue of layer gaps in a 3D fused filament modeling (FDM) printer, Takagishi and Umezu developed 3D-chemical high-temperature finishing (3D-CMF), which utilizes a pen-shaped device packed with a chemical substance capable of dissolving the materials needed to form structures produced in 3D. By dissolving the convex regions of the layer grooves on the structure's surface and filling the concave portions, this approach guarantees the cleanliness, selectivity, and durability of the 3D printed construction [13].

In order to initiate in-plane crystallization, current photolithography equipment requires continuous spectrum. Nevertheless, monochromatic light exposure over time removes the polymerization constraint. The thickness limitation and the need for rapid printing speeds in three-color (UV, red, and blue light) direct recording photolithography were both reduced thanks to the high converting of blue light (without UV light), which was bolstered by pre-initial irradiation with red light [14].

Thin shells that lack the volume to withstand or absorb impacts from changes in features are severely compromised by the mesostructure of 3D printing materials and inherent voids. Despite this, they have many applications, particularly in the aerospace and structural design industries because to their high strength-to-weight ratio. Prior FEA approaches for thin-walled buckling difficulties have anticipated that the covering is uniform and free of flaws; however, this is not the case for 3D printing processes described in commonly used materials (acrylonitrile-butadiene-styrene (ABS) and polylactic acid) [15].

The surface as well as the structural and geometrical characteristics of ABS are influenced by a thermal process known ironing, which allows FDM ABS products to have almost twice the deflection at the break and improved maximum tensile strength. It is possible to reduce the Ra factor of ABS parts by as much as 60% and the distortion by as much as 33% [16].

From a recycling standpoint, accessible 3D printing technology may improve the quality of polymer materials' primary (1), secondary (2), tertiary (3), and quaternary (4) processing [17]. For additive manufacturing (AM) applications using polymer materials with various reinforcements, this is particularly the case.

These images show the results of some of the most common problems with 3D printing, such as stringing, leaking, whiskers, hairy prints, and others.

Over- or under-extrusion occurs when the printer produces too little or too much of the desired material. Insufficient infill is produced when the print speed is too high defects on the manufactured object's surface, such as scars, scratches, or drips, caused by incorrect settings, such as temperature, flow rate, printing speed, or a flawed slicer profile. There is insufficient bonding between the exterior and interior layers.

When the extruder or bed moves during printing and restarts at a new location on the build plate, this is known as a layer shift or a shift in the X or Y axis.

The printing process is paused or filament is not being supplied correctly. The aforementioned wide variety of problems might benefit from more accurate AI-controlled 3D printing methods.

The automatic computer simulation of internet backbone data in real or near real time approach for 3D printing possibilities also provides a standard the configuration, manufacturing, able to monitor of condition monitoring, and management of the entire life cycle of 3D printed items, giving concrete form to the Internet of Things (IoT) belief system (layers: things-network-cloud) and Industry 4.0 (a plethora of inter-connected computing, information exchange, and storage systems).

By allowing for optimal parameter selection, ANNs serve as an optimization and prediction system that improves the efficiency of 3D printing plans. The ANNs used to study, analyze, and forecast the relationships between these pairs of variables account for the non-linear nature of these relationships.

Details about them that are not readily apparent in more conventional analyses. Bandwidth utilization is unaffected by the existence of complex, incomplete, or noisy datasets due to the fact that similarities in the measured datasets reinforce the interconnections between the synapses in the successive ANN layers (input, hidden, output). The function generator of the hidden layer is used in conjunction with the data input to calculate a response (estimated) that is more accurate, sensitive, and specific than would be possible using traditional statistical techniques alone. The new optimization decision criteria are compatible with the vast majority of 3D printing materials, processes, and characteristics, and the networks learn to accommodate novel criteria as they emerge. By insinuating and anticipating latent causal relationships between massive collections of characteristics, it is possible to provide optimal and original answers to technological issues.

The MLP-5-17-9 (Table 13.1) is a very simple ANN based on multilayer perceptron, and it has been shown to be effective at solving problems similar to those described, achieving (after 1000 epochs):

- Quality (learning): 0.8911; MSE (for training set data): 0.02.
- A quality score of 0.9134 was obtained during testing.

Data sets were gathered from commercial and academic 3D printing applications and used to train machine learning. In the beginning, the

**Table 13.1** Characteristics of used inputs and outputs.

| Input data | Output data | Source of datasets |
|---|---|---|
| Maximum data input | What is the internal temperature what is the manufacturing temperatures what is the second nozzles temperatures what is the fill volume fraction what is the print speed what is the layer height what is | Extractions of data from the 3D printer's software |

datasets were divided into three parts: learning (70%) testing (20%) and validation (10%).

During the preprocessing phase, the input parameters were altered such that they were all within the same range of values. Also, the network weights were predetermined, standardized, and initially set to values between $^{-}1$ and 1. This prevented the network's relative weights from being unfairly biased against the goal.

Using the Statistical and Pattern Recognition Toolbox and the Computational Intelligence Toolbox in MATLAB 16.0 (MathWorks), the proposed ANN-based solution was developed, analyzed, and refined. The architecture of the ANN was optimized with the help of a genetic algorithm (GA). More neurons were needed in the hidden component of the network to accommodate the complex interactions to the activation function. Our results suggest that MAXNET ANN could also provide equivalent results.

### 13.4.3 Choose High-Performance Material

Tuning all process conditions is necessary for the processing of novel, high-performance materials. Here, many sensors are used to keep an eye on the 3D printing process. We evaluate the data flow using AI to unearth previously unseen connections. Where artificial intelligence shines is in its ability to swiftly handle massive volumes of data, a task that would otherwise be impossible for humans. This research has allowed scientists to develop methods for processing complicated alloys while maintaining their original characteristics. CI's vision system, which it employs in its machine learning operations, can mix several (up to ten) materials.

An innovative approach to fabricating stretchable and scalable organic thermoelectrics is via the use of printed thermoelectric polyurethane magnetic nanoparticles reinforced with carbon nanostructures. None of the

3D-printed samples varied in electrical resistivity or thermoelectric factor in either the print or cross-print directions. 3D printed thermoelectric generators (TEG) using TPU/MWCNT as the organic thermoelectric material have the potential to be used in large-scale energy storage applications [18].

Yet, advances in 3D bio-indication techniques have opened up exciting new avenues for the design and production of bone implants for patients, negating many of the drawbacks of traditional ways of treating bone abnormalities. Composite materials, such as bioceramics, may have a significant impact on *in vivo* media's resistance to polymer breakdown. 3D printing of bone might benefit from a new nano-architecture of hydroxyapatite (HA) in a polymerization filler material like PLA [19].

Similar challenges may be solved using the ANN (Table 13.2), a reasonably basic ANN based on a multilayer perceptron, which has been shown to achieve (after 1000 epochs):

Quality (learning) = 0.9143; MSE = 0.001 for training set data.
The Quality of Our Tests is 0.9303.

Even though it may be necessary to add sets of standardized tasks in the form of printing sample parts in later stages of investigation, we based our study on operational data acquired from 3D printers in the framework of both academic and business investigations. Seventy percent was designated for teaching, twenty percent for testing, and ten percent for reviewing (validation). The control variables were parameterized to have the same highest and lowest points, and the weights were approximately and randomly determined between −1 and 1 at network initialization to remove the risk of bias.

Statistics and Pattern Recognition Toolbox and Deep Learning Toolbox were used in MATLAB 16.0 (MathWorks) for the ANN construction. The architecture of the ANN was optimized with the help of a genetic algorithm (GA). Our results suggest that MAXNET ANN could also provide equivalent results.

Plastic, the most common material used in 3D printing, is not biodegradable and has a low recycling capacity, both of which contribute

**Table 13.2** Best material strength.

| **Input data** | **Output data** | **Source of data sets** |
|---|---|---|
| Best dimension of inputs | Format recommendation | Directly taken datasets directly from the 3D printer's program |

considerably to environmental contamination. The broad use of 3D plastic digital printing necessitates the incorporation of sustainable practices, such as the reduction of waste, the minimization of energy usage, the safe gathering and reuse of trash, and the lifecycle monitoring and recycling of 3D printed items. Products that are 3D printed with more attention to their materials and quality may better meet the needs of their users, be used for longer, have their life cycles handled more effectively, and ultimately be recycled. Fast computation speeds, good quality, and low MSE are just a few of the ways in which the study's results facilitate prediction and evaluation of attribute and material selection. In turn, this makes it easier to perfect the procedure for 3D printing in the real world. In-depth research into the processes involved in the life cycle and reuse of substances used in 3D printers is vital, and the abovementioned AI-based solutions are only the beginnings. The processes of oversight and improvement of 3D printing goods are vulnerable to multidimensional evaluation and so need for a comprehensive approach, prompt diagnoses, surveillance, and efficient remedies.

### 13.4.4 3D Printing Problem Solving

Pattern recognition problem-solving and elimination-based problem-solving may have unrealized promise. Our mission is to compile a database of the most common 3D printing problems, together with autonomous CI-based solutions to those problems (helpdesk bot). There are three main areas where the indirect, unrealized potential of CI in 3D printing may be seen:

- Bioinspired labs-on-a-chip and connected systems [20–29]; the use of biodegradable polymers for teaching, testing (neuro) surgical access paths or implants, and strengthening damaged tissue with pharmaceuticals; and bio-ink for printing sensory nerve and lymph vessels skin, organ parts like the heart [22, 23], and muscles [24, 25].
- Verbal exchanges (including the maturing abilities required by the Internet of Things and Industry 4.0, which are being enhanced by AI) [30–32].

Even state-of-the-art technologies like iris, face, and fingerprint image identification may not be adequate to provide data safety [33] when pitted against a 3D reproduction of features generated by medical photography and 3D printing.

Sustainable printing processes might be aided by CI-based technologies throughout the lifetime of 3D printers.

## 13.5 ANN Methods

Literature examples and our own ANN-based methods demonstrate the feasibility of CI-based problem-solving in 3D printing. Although though CI-based decision support is not widely used today, this should change as the complexities of 3D printing and associated techniques and materials continues to increase.

The proposed AI methods, albeit simple in design, are effective and may make significant contributions to our understanding of 3D printers' material and characteristic selections and to their fully or remotely controlled improvement. It is important to implement the above- mentioned features and functionality into Industry 4.0 systems involving 3D printing in order to effectively come up with a strategy for optimizing material properties, life cycle control, general environmental evaluation and monitoring, and avoiding the ecological risks related to commercial 3D printing and produced environmental damage.

The provided methods are helpful because they have a small mean squared error (MSE), high quality, require little time to calculate, and may be easily implemented in real-time systems (0.01–0.02 for both). They have made some progress toward their ultimate aim of enhanced product customization.

While the efficacy of the CI-based technology is not in question, it is not intended to replace current 3D printing technology but rather to augment it in the areas of substance and features selections. This requires not only state-of-the-art 3D printing gear and programming but also the development of novel substances with better characteristics and growth temperature for operating machines.

Standardized methods and standards have been codified to streamline the process of recording and passing on knowledge to future generations of epidemiologists, doctors, and trainees. Technology toward the end of the 20th century, such as computerized datasets, ICT for transmitting information, 3D printing, laser scanners, effectors, CI algorithms, systems for decision-making support, etc., has become crucial to the inventive operations, analyses, and exchanging of expertise and understanding. Also, technology has begun to impact choices and behaviors (within business intelligence and strategic planning). Modern technological advancements like CI will usher in a new age of unprecedented possibilities.

Moreover, organizational, technological, and material issues should be considered as future research directions for CI-based 3D printing and reverse engineering assistance.

Expert deficiency and unequal development in the several fields of advanced technology that enable 3D printing—including materials science, nanostructures, and atom technology—could slow down future advancement in areas like enhanced diagnostic devices, neuro-prostheses, and brain-computer interfaces. Expanding the use of digital technologies, particularly better data processing, interpretation, and predictions, is necessary to break through some of these roadblocks and aid in the development of the next breakthroughs.

## 13.6 Conclusions

The integration of computational intelligence-based image classification techniques in the context of 3D printing within Industry 4.0 holds tremendous potential for revolutionizing the manufacturing processes. This abstract has explored the applications and benefits of utilizing computational intelligence, including machine learning, neural networks, and evolutionary algorithms, for image classification in 3D printing. Traditional image classification methods may struggle to accurately interpret complex designs and intricate details required for successful 3D printing. However, by leveraging computational intelligence techniques, manufacturers can achieve higher accuracy and efficiency in the classification process. Machine learning algorithms enable models to learn from large datasets, improving precision and reliability in classifying images for 3D printing. Additionally, neural networks provide powerful tools for recognizing and classifying intricate patterns and features crucial for accurate printing. The integration of evolutionary algorithms further enhances the benefits of computational intelligence in 3D printing. These algorithms optimize design and fabrication processes, leading to improved print quality, reduced material waste, and increased overall productivity. By exploring a range of design parameters, evolutionary algorithms identify the most efficient configurations, contributing to smarter manufacturing systems and advancing the goals of Industry 4.0. Integrating computational intelligence-based image classification in 3D printing within Industry 4.0 offers numerous advantages. It enables automated decision-making, reducing human intervention and potential errors. Real-time monitoring and adaptive control of the printing process enhance productivity and responsiveness. These techniques contribute to the development of intelligent and adaptive manufacturing systems, making the manufacturing process more efficient, precise, and cost-effective. However, challenges such as data quality, scalability, and interpretability must be addressed for successful implementation. Ensuring the availability of high-quality training data, scalability for large-scale manufacturing, and interpretability of classification results are essential considerations. In conclusion, the integration of computational intelligence-based image classification techniques in 3D printing within the framework of Industry 4.0 has the potential to revolutionize manufacturing processes. By leveraging machine learning, neural networks, and evolutionary algorithms, manufacturers can achieve higher accuracy, efficiency, and productivity in 3D printing. Continued research and development in this field will likely lead to transformative applications, paving the way for smarter, more adaptive, and sustainable manufacturing systems in the future.

## References


1. Lu, T., Towards a fully automated 3D printability checker, in: *Proceedings of IEEE International Conference Industrial Technology*, vol. 2016, pp. 922–927, 2016.
2. Mintz, Y. and Brodie, R., Introduction to artificial intelligence in medicine. *Minim. Invasive Ther. Allied Technol.*, 28, 2, 73–81, 2019.
3. Paraskevoudis, K., Karayannis, P., Koumoulos, E.P., Realtime 3D printing remote defect detection (stringing) with computer vision and artificial intelligence. *Processes*, 8, 1464, 2020.
4. Esteva, A., Kuprel, B., Novoa, R.A. *et al.*, Dermatologist-level classification of skin cancer with deep neural networks. *Nature*, 542, 7639, 115–118, 2017.
5. Wu, Y., Peng, G., Chen, L., Zhang, H., Service architecture and evaluation model of distributed 3D printing based on cloud manufacturing. *2016 IEEE International Conference on Systems, Man, and Cybernetics (SMC)*, pp. 002762–002767, 2016.
6. Minetto, R., Volpato, N., Stolfi, J., Gregori, R.M.M.H., da Silva, M.V.G., An optimal algorithm for 3D triangle mesh slicing and loop-closure. *Comput.-Aided Des.*, 92, 1–10, 2017.
7. Gregori, R.M.M.H., Volpato, N., Minetto, R., da Silva, M.V.G., Slicing triangle meshes: An asymptotically optimal algorithm, in: *Proc. 14th International Conference on Computational Science and Its Applications*, pp. 252–255, 2014.
8. Liao, H., Mesfin, A., Luo, J., Joint vertebrae identification and localization in spinal CT images by combining short- and long-range contextual information. *IEEE Trans. Med. Imaging*, 37, 5, 1266–1275, 2018.

9. Sun, L., Wang, J., Hu, Z. *et al.*, Multi-view convolutional neural networks for mammographic image classification. *IEEE Access*, 7, 1–10, 2019.
10. Dhatterwal, J.S., Kaswan, K.S., Ojha, R.P., The role of multiagent system in industry 4.0, in: *A Roadmap for Enabling Industry 4.0 by Artificial Intelligence*, pp. 227–246, 2022.
11. Ligon, S.C. *et al.*, Polymers for 3D printing and customized additive manufacturing. *Chem. Rev.*, 117, 15, 10212–10290, 2017.
12. Bourell, D. *et al.*, Materials for additive manufacturing. *CIRP Ann. Manuf. Technol.*, 66, 659–681, 2017.
13. Takagishi, K. and Umezu, S., Development of the improving process for the 3D printed structure. *Sci. Rep.*, 7, 39852, 2017.
14. Attaran, M., The rise of 3-D printing: The advantages of additive manufacturing over traditional manufacturing. *Bus. Horiz.*, 60, 677–688, 2017.
15. Tofail, S. *et al.*, Additive manufacturing: Scientific and technological challenges, market uptake and opportunities. *Mater. Today*, 21, 1, 22–37, 2018.
16. Kaswan, K.S., Dhatterwal, J.S., Sood, K., Balusamy, B., Role of blockchain technology in the modern era, in: *Blockchain Technology in Corporate Governance: Transforming Business and Industries*, pp. 1–28, 2022.
17. Adamczak, S., Bochnia, J., Kaczmarska, B., Estimating the uncertainty of tensile strength measurement for a photocured material produced by additive manufacturing. *Metrol. Meas. Syst.*, 21, 553–560, 2014. https://doi.org/10.2478/mms-2014-0047.
18. Ahlers, D., Wasserfall, F., Hendrich, N., Zhang, J., 3D printing of nonplanar layers for smooth surface generation, in: *IEEE International Conference Automative Science Enginerring*, 2019, https://doi.org/10.1109/COASE.2019.8843116.
19. Najafi Zare, R., Doustkhah, E., Assadi, M.H.N., Threedimensional bone printing using hydroxyapatite-PLA composite. *Mater. Today Proc.*, 42, 1531–1533, 2021.
20. Dhatterwal, J.S., Kaswan, K.S., Preety, D., Balusamy, B., Emerging technologies in the insurance market, in: *Big Data Analytics in the Insurance Market*, pp. 275–286, 2022.
21. Aoyagi, K., Wang, H., Sudo, H., Chiba, A., Simple method to construct process maps for additive manufacturing using a support vector machine. *Addit. Manuf.*, 27, 353–362, 2019. https://doi.org/10.1016/J.ADDMA.2019.03.013.
22. Beaman, J.J., Bourell, D.L., Seepersad, C.C., Kovar, D., Additive manufacturing review: Early past to current practice. *J. Manuf. Sci. Eng. Trans. ASME*, 9, 1–24, 2020. https://doi.org/10.1115/1.4048193/1086507.
23. Caggiano, A., Zhang, J., Alfieri, V., Caiazzo, F., Gao, R., Teti, R., Machine learning-based image processing for on-line defect recognition in additive manufacturing. *CIRP Ann.*, 68, 451–454, 2019. https://doi.org/10.1016/J.CIRP.2019.03.021.

24. Peele, B.N., Wallin, T.J., Zhao, H., Shepherd, R.F., 3D printing antagonistic systems of artificial muscle using projection stereolithography. *Bioinspir. Biomim.*, 10, 5, 055003, 2015.
25. Harren, K., Dittrich, F., Reinecke, F., Jäger, M., Digitalization and artificial intelligence in orthopedics and traumatology. *Orthopade*, 47, 12, 1039–1054, 2018.
26. Dhatterwal, J.S., Kaswan, K.S., Baliyan, A., Jain, V., Integration of cloud and IoT for smart e-healthcare, in: *Connected e-Health: Integrated IoT and Cloud Computing*, pp. 1–31, Springer International Publishing, Cham, 2022.
27. Preety, K.S.K. and Dhatterwal, J.S., Fog or edge-based multimedia data computing and storage policies, in: *Recent advances in Multimedia Computing System and Virtual Reality*, p. 239, Routledge Taylor & Francis Group, Milton Park, Abingdon in the United Kingdom, 2021.
28. Kokkinis, D., Bouville, F., Studart, A.R., 3D printing of materials with tunable failure via bioinspired mechanical gradients. *Adv. Mater.*, 30, 19, e1705808, 2018.
29. Tawk, C., In Het Panhuis, M., Spinks, G.M., Alici, G., Bioinspired 3D printable soft vacuum actuators for locomotion robots, grippers and artificial muscles. *Soft Robot.*, 5, 6, 685–694, 2018.
30. Khadilkar, A., Wang, J., Rai, R., Deep learning–based stress prediction for bottom-up SLA 3D printing process. *Int. J. Adv. Manuf. Technol.*, 102, 5, 2555–2569, 2019.
31. Tlija, A., Wegrzyn-Wolska, K., Istrate, D., Missing-data imputation using wearable sensors in heart rate variability. *Bull. Polish Acad. Sci. Tech. Sci.*, 68, 2, 255–261, 2020.
32. Szczepanski, J., Klamka, J., Wegrzyn-Wolska, K.M., Rojek, I., Prokopowicz, P., Computational intelligence and optimization techniques in communications and control. *Bull. Polish Acad. Sci. Tech. Sci.*, 68, 2, 181–184, 2020.
33. Huang, J. and Rai, R., Conceptual three-dimensional modeling using intuitive gesture-based mid-air three-dimensional sketching technique. *J. Comput. Inf. Sci. Eng.*, 18, 4, 1–13, 2018.

# 14

# Bioprinting and Robotics Engineering: Applications, Recent Progress, and Future Directions

**Pawan Whig[1]*, Shama Kouser[2], Ashima Bhatnagar Bhatia[3], Rahul Reddy Nadikattu[4] and Yusuf Jibrin Alkali[5]**

[1]*Vivekananda Institute of Professional Studies, New Delhi, India*
[2]*Department of Computer Science, Jazan University, Jazan, Saudi Arabia*
[3]*Vivekananda Institute of Professional Studies, New Delhi, India*
[4]*University of the Cumberland, Kentucky, USA*
[5]*Federal Inland Revenue Service, Abuja, Nigeria*

***Abstract***

Bioprinting technology has the potential to be used in synthetic biology for the reproducible andprecise production of scaffolds, cells, tissues, and organs. Based on their basic printing principles,bioprinting technologies are broadly classified into three categories: inkjet-based bioprinting, pressure-assisted bioprinting, and laser-assisted bioprinting. Each of these printing processes has advantages and disadvantages. Bioprinting makes use of biomaterials, cells, or cell factors as "bioink" to create potential tissue architectures. The printed result is heavily influenced by biomaterial factors such as biocompatibility,cell viability, and the cellular microenvironment. Various printing technologies have been researched, withsignificant progress made in printing various types of tissue such as vasculature, heart, bone, cartilage, skin,and liver. This overview explains the fundamental concepts and major features of certain commonly used printing technologies. We concentrate on recent developments in three-dimensional printing applications, as well as present issues and future directions. Also a case study will be discussed at the end of the chapter which will be very useful for the researchers working in the same field.

***Keywords:*** Bioprinting, biolink, cells, tissues, inkjet, microenvironment

**Corresponding author*: pawanwhig@gmail.com

E. Gangadevi, M. Lawanya Shri, Rajesh Kumar Dhanaraj and Balamurugan Balusamy (eds.) Computational Intelligence in Bioprinting, (275–302) © 2024 Scrivener Publishing LLC

## 14.1 Introduction

In today's environment, 3D printing and robotics are not unconnected. We are all aware that the concepts of robotics form the basis of 3D printing machines. However, as technology has advanced, 3D printing is poised to "give back" to its underlying technology - by contributing in the creation of robots [1]. In popular culture, we've seen mad robots like Terminator, Wall-E, and iRobot. In reality, robots are not always as bizarre. A slew of issues must be considered, beginning with a goal that is a little more practical than assisting in the rescue of the world from an alien dictator. Mechanical engineering, electronics, biotechnology, computer science, and other disciplines contribute significantly to the construction of a robot.

One of the robotics applications is 3D printing, which automates the most arduous and repetitive operations while lowering costs and increasing turnaround time. Simultaneously, 3D printing enables the rapid production of many types of robots and their components [2]. Today, 3D printing is performed both 'for' and 'by' Robotics.

Bioprinting is expected to replace traditional production processes that take hours to finish a task. The advancement of 3D printing allows it to permeate industries such as automobiles, aircraft, and healthcare. We mostly utilize robots for activities that are unfeasible, dangerous, expensive, repetitive, or time-consuming for people [3]. Such as inspecting radioactive sites, defusing bombs, and investigating severe temperatures or pressures. A flowchart of bioprinting techniques is shown in Figure 14.1.

Across history, many scientists, inventors, intellectuals, and others have thought that robots will one day be able to emulate human behaviors [4]. While this is not yet true in its entirety, technology is evolving at such a rapid pace that a fully intelligent humanoid may not be far distant.

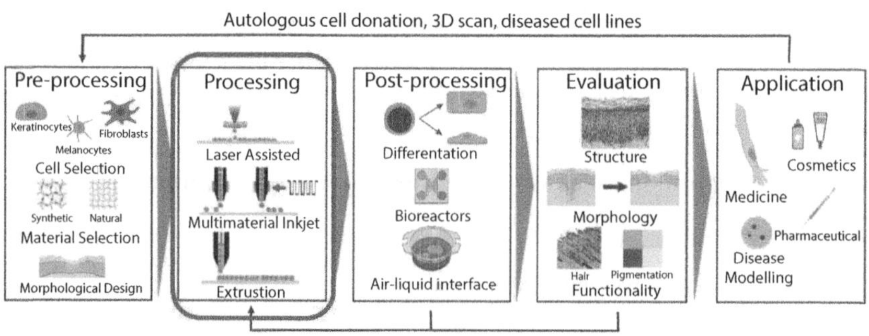


**Figure 14.1** Flowchart of bioprinting techniques.

Approximately 20 years ago, practically all of the robots in existence could be found at automobile assembly plants. Mechanical arms are typically used to weld vehicle parts. You may now have a humanoid butler robot bring you coffee in the morning [5].

## 14.2 Background

Robotic machines and automata have been described for at least a couple of thousand years. Heron of Alexandria's Pneumatical and Automate contain over 100 descriptions of fire engines, a coin-operated machine, a wind instrument, and other devices. Flash forward nearly 1500 years, and Leonardo da Vinci was tinkering with a robotic monk with lifelike foot, eye, lip, and head motions. This was the first tangible manifestation of a robot that looked like a human being [6].

The industrialization in the previous few centuries, as well as the digital revolution much more recently, have stimulated advances in this subject. It's been a long road from Nikola Tesla's radio-controlled warships in 1898 to self-navigating Autonomous Indoor Vehicles (AIVs) inside Tesla's new factory in California [7].

Technology is gradually enabling the quick creation of three-dimensional items made of plastic, nylon, and metal. While this may spark a manufacturing revolution in and of itself, the recent invention of bioprinters is even more exciting. Living tissues are artificially produced in a layer-upon-layer manner using these bioprinters. The merging of these small layers forms an organic entity over a period of several hours. The bioprinter extrudes a hydrogel alongside the cells to provide cell support and safety. 3D printing machine have received a lot of interest in recent studies because of their benefits over traditional lab-grown organ growth, among other things. One printing machine may produce several print models without requiring any changes to the printer's core mechanical or electrical components as shown in Figure 14.2. The most significant benefit of 3D bioprinting is personalization and versatility. It increases production and efficiency while also hastening medical operations such as kidney transplants [8].

Three-dimensional bioprinters are developed with numerous qualities that distinguish them from traditional 3D printers. These characteristics encompass biological processes such as cell development and tissue formation. Cell structures may be printed on petri dishes or molds using some bioprinters. This simplifies the handling of structures in post-printing operations [9]. To achieve precise printing of structures according to unique requirements, high precision motors are necessary. Most significantly,

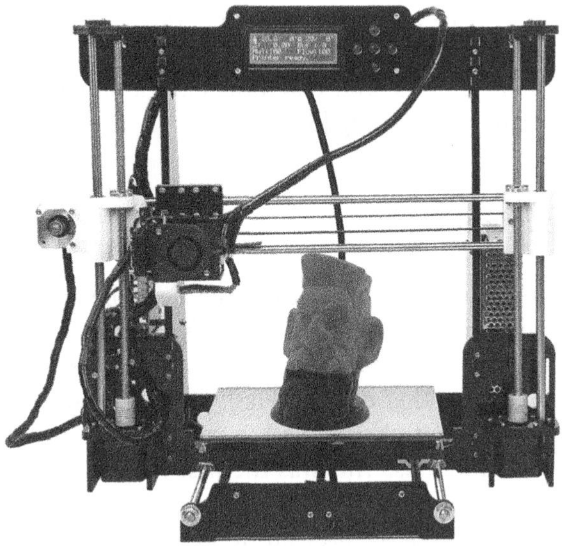

**Figure 14.2** 3D printing machine.

to minimize cell contamination, bioprinting must be performed in a sterile environment. GeSim's Bioscaffolder, Inkredible+, and NovoGen MMX are a few commercial 3D bioprinters that have arrived on the market.

Despite being still in the early stages of development, 3D bioprinters have the potential to transform medical practice. Bioprinters distribute cells from a print head that moves in three dimensions to deposit the cells exactly where they are needed. Bioprinters, on the other hand, may be built in a variety of ways. Organovo, for example, is one of the top ten 3D bioprinters that employs a robotic arm and can execute the majority of the duties that a linear 3D bioprinter can. Because of its small size, it can reach regions that linear bioprinters cannot. Overall, the subject of bioprinting is constantly progressing, with new approaches being explored for extra features and improved printability [10].

Bioplastics are either compressed air or physically injected into the extruder. Printing with varied viscosities of bionics is possible using pneumatic-driven devices that use air pressure. However, due to the volume of compressed gas, these systems may have delays. Mechanical-driven systems, on the other hand, use screws or pistons to control the flow rate of bioink. They are, however, delicate and can trigger cell death if not constructed appropriately. Hydrogels, collagen, and gelatin are some of the several types of bioinks [11].

Ego ultrashort peptides with three to seven amino acids have been shown to function as natural yet synthetic 3D scaffolds that encapsulate 99.9% of water and compete well with current bioinks and hydrogels. As a result, they are an excellent choice for numerous organ construction and systems biology applications. Their capacity to assume shape and quickly be chemically changed gives them an edge in 3D bioprinting.

## 14.3 3D Printing

3D printing is a relatively recent manufacturing method that is mostly performed by robots or in a robotic manner. It is a fully automated process in which machines construct an item layer by layer. In less than two decades, 3D printing has moved from labs and a few specialized enterprises to the worldwide market, embracing industries ranging from toys to aircraft. Initially limited to a few plastic materials, 3D printing today supports a wide range of polymers, metals, and even composite materials such as Carbon Fiber [12].

The Industry 4.0 revolution highlights the importance of 3D printing technology as shown in Figure 14.3. According to experts, additive manufacturing techniques will become increasingly prevalent in a variety of businesses throughout the world in the next years [13].

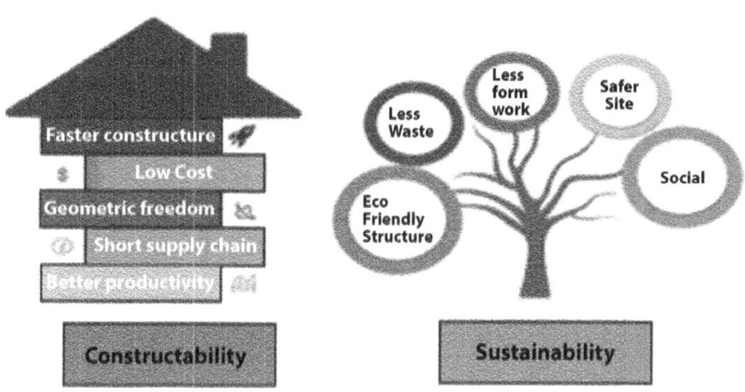


**Figure 14.3** Importance of 3D printing technology.

While 3D printing has been used in industry since the early 1980s, the entry cost for 3D printers has just recently decreased dramatically. In today's age of digital production, adopting 3D printing is an efficient and wise decision.

## 14.4 3D Printing Applications

When 3D printing initially hit the market, it was mostly utilized for prototyping. Because the resources available were restricted, firms were able to create low-cost plastic-based prototypes. However, with additional study and technical advancements in the technique and material selection, 3D printing will reach its peak of success in the near future. The chart below in the Figure 14.4 demonstrates the different applications of 3D printing in industries [14].

### 14.4.1 Tooling for Prototyping and End-Use Parts

The graphic shows that the uses of 3D printing are becoming increasingly diversified. The transition from prototype to more common applications suggests that 3D printing will continue to rise. The expansion is obvious as 3D printing becomes a critical component in industries such as aircraft, drones, automotive, medical, military, architecture and construction, and so on [15–20].

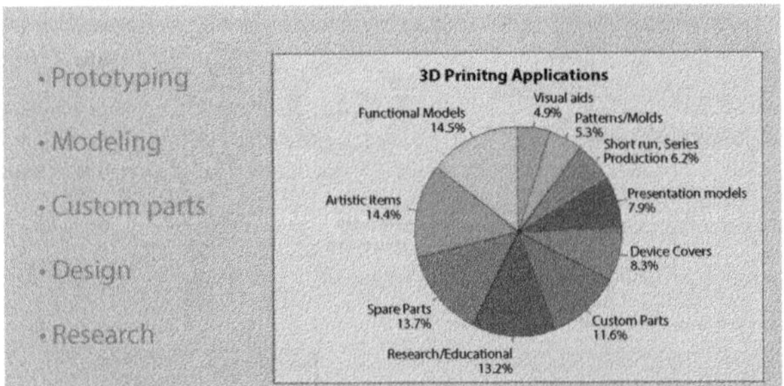


**Figure 14.4** Different applications of 3D printing in industries.

## 14.4.2 Industrial Uses of 3D Printing

### *14.4.2.1 Aviation*

Since 1989, the aerospace industry has used 3D printing. 3D printing has witnessed remarkable growth in this field in recent years. Some industrial applications of 3D printing are shown in Figure 14.5. The following are some significant aspects in the aircraft industries:

- Quick delivery
- Limited production quantity
- DfAM Possibility

These factors are critical in reducing the total weight of the aeroplane. This significantly reduces expenses while also preventing unnecessary CO2 emissions [21–25]. Here's how 3D printing is becoming an important part of the aircraft supply chain:

Design: The tools and technologies that power 3D printing provide designers a lot of leeway in trying out new ideas. The method of making components can range from 3D printed moulds to castings and direct printing. Designers have a lot of freedom when it comes to 3D printing. They can convert 3D CAD files to real printers and make revisions on the fly [26–30].

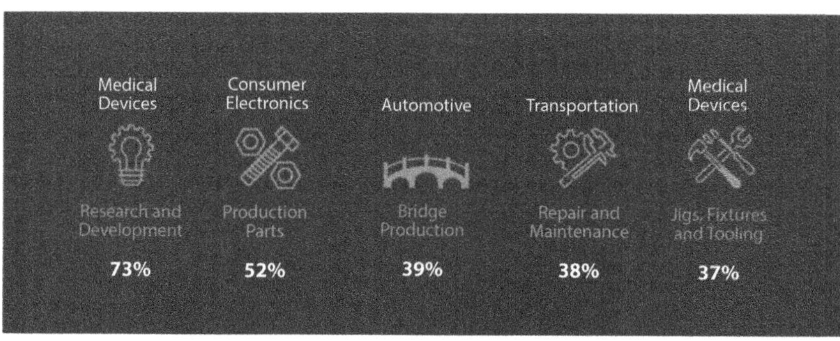


**Figure 14.5** Industrial applications of 3D printing.

Designing and Building: Design conceptualization is required to demonstrate the utility of particular models in aerospace production. Aerodynamic testing is also required. This is why high accuracy, working prototypes are required. Because of modern printing processes like FDM (Fusion Deposition Modeling), SLA (Stereolithography), and Material Jetting, 3D printing makes this work very simple.

Personalization: It is customary in aerospace to have customized parts for a specific design. Customization is simple, rapid, and powerful with 3D printing [31–35].

### 14.4.3 Which Are Some of the Most Typical Items Produced Using a 3D Printer?

Jigs and Fixtures: When compared to traditional production procedures, 3D printing of these components saves almost 60 to 90 percent on costs and improves lead time [36–39].

Surrogates: like prototypes, serve as placeholders that will be replaced during the final phases of assembly. Surrogates made from 3D printing are inexpensive.

Mounting brackets: with the advent of metal 3D printing, these steel installation brackets are now easily created.

### 14.4.4 Wing and Propellers

Airbus Services has built metal-printed flying replacement parts for an Airbus A320eco using additive manufacturing (3D printing). It is one of the oldest planes. This aircraft's wingtip fences are no longer manufactured. Airbus Services, on the other hand, can now build and deliver new replacement parts to some of the oldest aero planes using 3D printing [40–44].

3d printers is also used extensively in the construction of space rockets. Stargate is among the biggest metal 3D printers, and organizations such as Relativity Space utilize it to create a full rockets [45–49].

#### *14.4.4.1 Role of 3D Printing Automobile Industry*

3D printing is widely used in the automotive industry for prototype, tooling, and end-parts manufacturing. 3D printing and additive manufacturing have unlocked Pandora's Box. It is now feasible to create fresh styles as

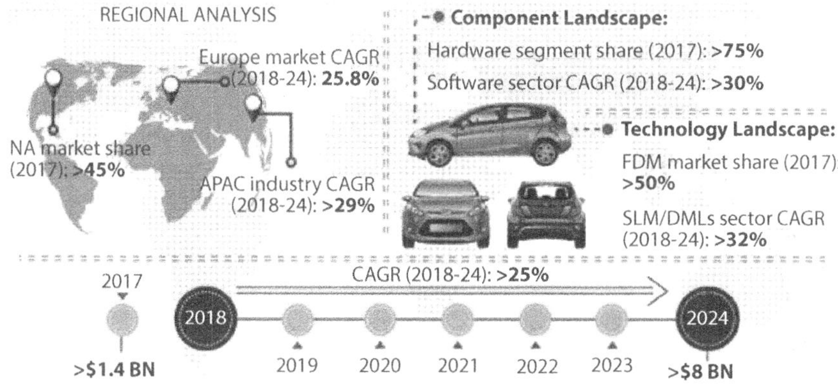


**Figure 14.6** Regional analysis report for role of 3D printing in automobile industry.

well as clean and lightweight items without sacrificing safety. 3D printing also has a shorter lead time and is less expensive [50–55]. In the automobile industry, 3D printing capabilities have evolved from quick prototyping to direct manufacture of final items. The regional analysis report for role of 3D printing Automobile Industry is shown in Figure 14.6.

### *14.4.4.2 Automobile Quality is Improved via Rapid Prototyping*

3D printing is ideal for prototyping. Rapid prototyping provides high-quality testing of final goods. The design freedom of 3D printing enables the creation of practical and long-lasting prototypes. General Motors creates prototypes of around 20,000 components using 3D printing processes such as SLS (selective laser sintering) and SLA (stereolithography) [56, 57].

### *14.4.4.3 Tool Customization*

Automotive tooling must be precise, consistent, and of high quality. Because of its versatility, 3D printing contributes to this. Customized tools may be created using 3D printing to increase labor and efficiency on the factory floor. BMW has used additive manufacturing (AM) or 3D printing to create custom-built, handcrafted tools that are 72% lighter than conventional equipment.

#### *14.4.4.4 Trying to Lose the Most Weight Possible*

Vehicle weight decrease correlates with vehicle fuel economy. Using lightweight components can help a lot in this endeavor. Many manufacturers have begun to replace aluminum body elements with carbon fiber-based materials. Ford's F-150 is ideal for the job. The vehicle's body was composed of aluminum in 2014. The weight was lowered by 317 kgs after it was replaced with carbon fiber.

## 14.5 Recent Progress in 3D Printing

### 14.5.1 Unmanned Aerial Vehicles (UAV)

Although officially a part of Aerospace, unmanned vehicles are becoming more popular due to their vast range of uses and may therefore be classified individually. Drones are utilized in construction, agriculture, photography, delivery, security, and other fields. This is why 3D printing techniques are being utilized to create personalized drones to meet the needs of the individual. All parts of the drones may be 3D printed and assembled. Propellers, antennae, blades, battery housing, landing gears, and other components may be simply 3D printed using the appropriate material.

### 14.5.2 Medicine

In the early 2000s, 3D printers made their debut in the medical industry. It was utilized in dentistry and personalized prostheses. As 3D printing has advanced, it is increasingly being employed in the medical profession. It can now create bones, exoskeletons, jawbones, cell cultures, ears, and other structures. Some of the most frequent applications for 3D printing in medicine are shown in Figure 14.7.

### 14.5.3 Organs and Tissues May Be Printed via Bioprinting

Organ procurement for persons with organ failure is a time-consuming process. Because there are fewer donors, there is a considerable scarcity of organs. 3D printing may currently be used to create organs or tissues using the following techniques:

- Beam
- Ink - jet
- Extruder

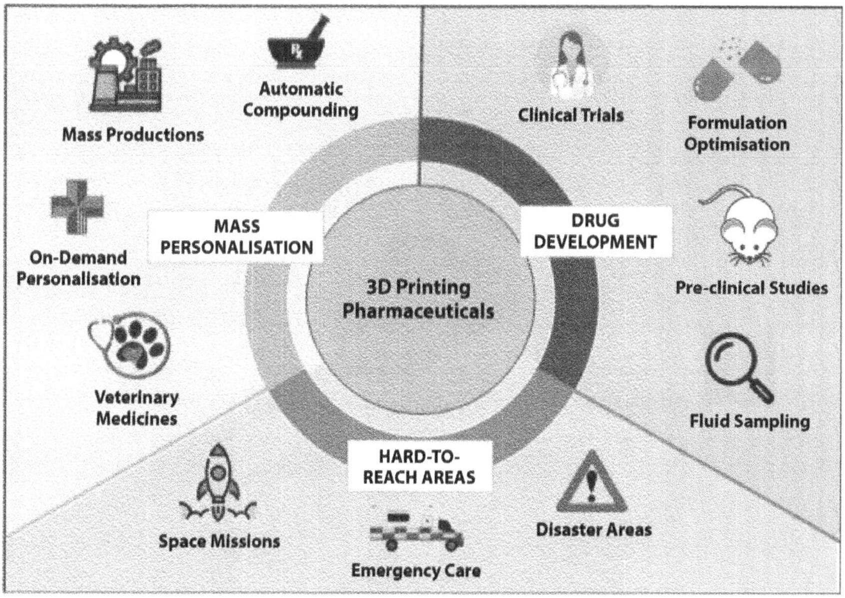


**Figure 14.7** Frequent applications for 3D printing.

The bio-ink is placed in droplets to mimic a live cell, which will aid in repair or reproduction.

### 14.5.4 Smart Building

In 2004, someone utilized 3D printing to construct a wall for the first time. A University of South Carolina professor attempted to build a wall using a 3D printer. This has resulted in the emergence of 3D printers in the building and architectural industries. The contour crafting method is a prominent approach in concrete 3D printing. A robotic arm glides back and forth, extruding successive layers of concrete or dirt. The material is oozed from the nozzle, and the silicone spatulas are used to smooth the layers. 3D printing role in Construction is shown in Figure 14.8.

### 14.5.5 Sand Dunes

A equipment named the "D-shaped 3D printer" was utilized by an Italian architect, Enrico Dini. The machine here employs a powder that gradually freezes into many layers that are applied one after the other. Similar to the SLS printing process. The machine can build buildings up to 6 cubic meters in size.

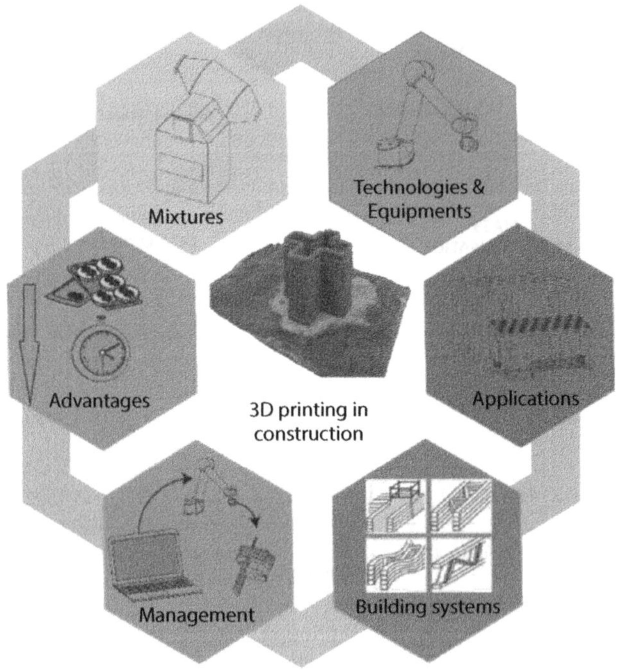


**Figure 14.8** 3D printing role in construction.

### 14.5.6 Metal Engraving

Metal constructions may now be 3D printed at a pace of 2 kilograms (material) per hour. MX3D, a Dutch business, invented this method. Wire Arc Additive Manufacturing is the name given to this metal printing method.

### 14.5.7 3D Printing's Use in Robot Construction

Manufacturing is broadening and penetrating new sectors as it becomes more digital and nimbler. Robotics is one area where advancements in additive manufacturing can have a significant impact on the industry's growth. Here are some of the benefits of additive manufacturing:

- Prototyping
- Production at a low cost
- Quick time to market
- Reduce your waste.

### 14.5.8 Characteristics

#### *14.5.8.1 Robots Are Not Like Other Industrial Items*

Robot manufacture is more customized and specialized. Robot design and functionality are frequently altered to meet a specific need. This is why 3D printing can help the robotics industry expedite manufacturing. There is greater flexibility for experimentation because prototypes are simple to produce and test.

#### *14.5.8.2 Design Versatility and Rapid Prototyping*

Design engineers may now swiftly build the tools and fixtures required for robot assembly using 3D printing. It also allows for the production of parts without the need for molds. Using sophisticated additive manufacturing processes, achieving complicated and unique shapes is already simplified. 3D printers are now more capable than ever of handling a wide range of materials.

#### *14.5.8.3 Lower Output Figures*

Despite their extensive use, certain types of robots are still quite rare in the globe. Because of the low number, it is unappealing to typical production procedures. Who would build a production line to make only 100 robots in a year? That is where 3D printing comes in: technology enables on-demand production while eliminating the upfront expense of establishing a facility, assembly line, and so forth.

#### *14.5.8.4 Robot Types and Applications*

Let's look at several different sorts of robots (in no particular order) and see how 3D printing and additive fabrication have aided their development. They labor in regulated conditions where no surprises are to be expected and serve a single goal every day. One such example is a robotic welding arm at an automobile industry. A humanoid robot may be constructed from 3D printers and utilized to execute 3D printing. It is now feasible to print from various angles and things with complicated geometry using a robotic arm.

Of course, 3D printing has facilitated rapid prototyping and iterative design in the development of such robotic hands. When complicated joints are included in such robotic hands, 3D printing is very advantageous in both the development and production stages.

Table 14.1 underscores the transformative potential of 3D printing in agriculture by emphasizing its adaptability, agility, and capacity to revolutionize how agricultural solutions are conceived and implemented.

**Table 14.1** Characteristics of 3D printings.

| Process | Materials | Principle | Advantages | Disadvantages |
|---|---|---|---|---|
| FFF | Thermoplatics | Fusion-driven restacking | Fast printing low cost<br>High strength | Pretreatment<br>Weak interaction<br>Rough surface |
| DIW | Thermoplastics<br>Elastomers<br>Hydrogels | Ink extrusion | Material diversity<br>Low cost | Nozzle clogging<br>Post-processing |
| DLP | Photosensitive resin | *In situ* photocuring | High resolution<br>Intricate fabrication | Weak strength<br>High cost |
| SLS | Thermoplastics | Laser sintering | Material diversity<br>High efficiency<br>High material utilization | Rough surface<br>Powder pollution<br>High cost |

#### *14.5.8.5 Independent*

They are programmed to deal with almost any event that may arise on their journey. They are frequently used in circumstances when human involvement would be preferable.

#### *14.5.8.6 Submarine Robot*

This remark emphasizes the importance of Deepwater exploration robots. Robots can be designed to travel to previously undiscovered areas of the ocean. Some of these underwater robots can be used to retrieve a map of the seafloor. Aside from that, a variety of near-surface applications necessitate the use of underwater robots, such as ship hull inspection and cleaning, underwater welding, underwater building inspection, and so on.

## 14.6 Future Directions in 3D Printing

A 3D-printed robot can be tailored to the ocean's environment and/or the specific sort of application. For example, if the work requires a certain type of welding torch to be handled, the holder or gripper may be created and manufactured in days rather than weeks/months. Furthermore, 3D printed parts may be made from a broad range of materials that are suited for specialized underwater applications such as seawater (salty), deep water, acidic/alkaline water, and so on. 3D printing materials, ranging from plastics to metals to composites, may now be used to create robots ranging from near-surface to deep ocean standards.

Despite the current boom in underwater robots, they are still constructed to order. Once again, 3D printing's adaptability and modularity are demonstrated.

### 14.6.1 Robots That Clean Solar Panels

Rooftop solar installation methods address one of our planet's most pressing issues: energy. However, we require a large amount of water to clean and maintain these solar panels. They are prone to a lot of dust since they are exposed to the open sky. Manual cleaning becomes tedious and moment. This technique may now be made more cost-effective by using robots as shown in Figure 14.9.

Some AI-enabled bots can wash every nook and crevice of these solar panels with minimum human interaction and low water use. Furthermore, the fact that 3D printing can produce any design in a matter of days makes it a preferred technique of fabrication for solar panel cleaning robots.

**Figure 14.9** Robots that clean solar panels.

The design may be tailored to the specs of the solar panel. The material selection for 3D printing allows it to be functional in a wide range of weather conditions for an open-to-weather condition such as solar panels. Solar panel robots may also benefit from a lot of automation because they operate on a large scale and usually on a schedule. Because 3D printing makes designing and integrating new components as simple as playing computer games, these robots can be regularly updated with new sensing and components. Figure 14.10 provides a visionary glimpse into the manifold applications of 3D printing poised to reshape the agricultural landscape. As technological advancements continue to unfold, these emerging possibilities hold the potential to revolutionize various facets of agriculture.

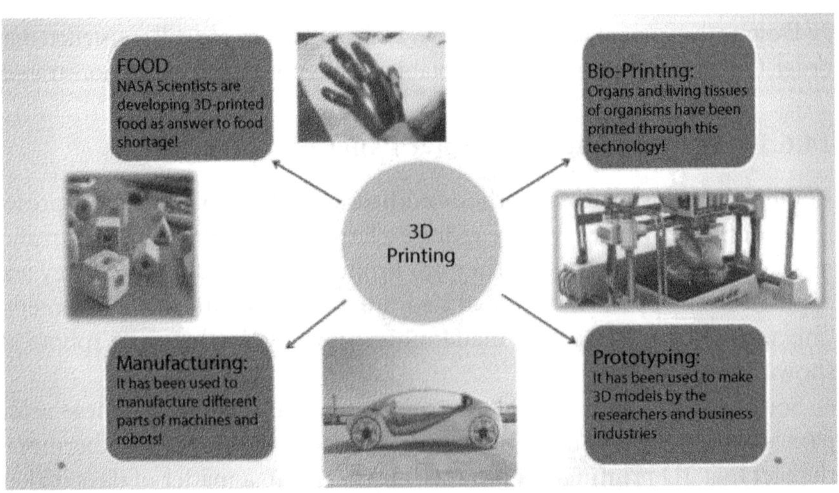


**Figure 14.10** Future applications of 3D printing.

### 14.6.2 Factory Robots

Robots have just joined the warehouse market and can transport, manage, and expedite the whole warehousing operation. Some corporations, such as Amazon and Ocado, are already using warehouse robots as shown in Figure 14.11. The manufacture of a wide range can benefit from warehouse robots:

- Automotive Electronics and Electrical
- E-commerce Retail Food & Beverage
- Chemical and metal industries
- Textile

There really are various ways in which 3D printing has aided the development of warehouse robots.

Many warehouse robot parts are currently manufactured utilizing additive manufacturing techniques. It is now feasible to make warehouses more agile and efficient utilizing 3D printed elements.

Warehouse robots are critical to the automobile sector. They are utilized in inventory storage for spare parts that are too heavy to lift. Robots can now do all of the heavy lifting. Wear and tear is not uncommon in warehouse robots due to their lengthy working hours. You can easily manufacture any damaged part and bring the robot back to function thanks to 3D printing. The warehouse does not need to retain a large inventory of robot parts.

Again, the adaptability and speed of 3D printing allow you to constantly improving these robots and making them adaptive to any "warehouse weather." Humanoids are robots that are designed to be as human-like as

**Figure 14.11** Factory robots used in various industries.

feasible. Humanoids are distinguished not just by their look but also by their functioning.

Complicated joints and pieces may be digitally manufactured using 3D printing. This provides engineers design freedom. Engineers, for example, now employ 3D modelling techniques to simulate the ball-joint movement of the human body. This saves a lot of time and makes modifications simple. When compared to their traditional equivalents, 3D printed humanoid components are lightweight and inexpensive.

Supplementing robots are designed to either augment or replace a person's abilities. Augmenting robots are frequently used in medical as prosthesis for missing limbs. Modern prostheses can learn to respond to neurological impulses attached to a working muscle, allowing the patient to move the prosthesis by moving the working muscle.

- 3D printing is especially special for prostheses that need to match the user's body shape. E.g. If it is a knee or leg support device, 3D printing can help create the prosthesis to the user's specific height, leg form, leg girth, hip shape, and so on.
- 3D printers are utilized to create lightweight prosthetic components. This is significant because a prosthesis should not be so heavy that it becomes a strain on the user's body.
- Open source contribution — People all over the world may now build fantastic 3D printed designs for prostheses that they feel would aid people suffering from various conditions. This removes the monopoly, and hence the high costs, of items developed by a single business.

Soft robotics is concerned with the construction of robots out of materials that are comparable to live beings. They attempt to mimic the behaviors of live organisms and adapt to their surroundings. Memory coils and actuators are used to mimic the behaviors of tendons and muscles.

### 14.6.3 Case Study

In this case study, we will do a basic linear regression evaluation on the 3D printing dataset to investigate the numerous associations that exist between the target variables, also known as labels, and the predictor variables, also known as features.

### 14.6.4 Adding the Dataset to the Notebook

We will import the 3D printer for mechanical engineers in this part.

csv file into the worksheet and examine some basic information like as the number of rows and columns, datatypes, comments (if any), and missing values (if any).

Before importing the dataset, import the machine learning libraries pandas, NumPy, matplotlib, and seaborn.

The dataset contains 12 columns.

The first 9 columns i.e from layer_height to fan_speed are features.

The last 3 columns i.e from roughness to elongation are labels. So, we will be predicting these three based on the 9 features.

The various units of the nine features are as follows:

1. Layer Height in mm
2. Wall Thickness in mm
3. Infill Density in %
4. Infill Pattern in either Grid or Honeycomb
5. Nozzle Temperature in Degree C
6. Bed Temperature in degree C
7. Print speed in mm/s
8. Material in either abs or pla
9. Fan Speed in %

Pictures of Infill patterns and Filament materials as shown in Figure 14.12 and structure from 3D Printing is shown in Figure 14.13.

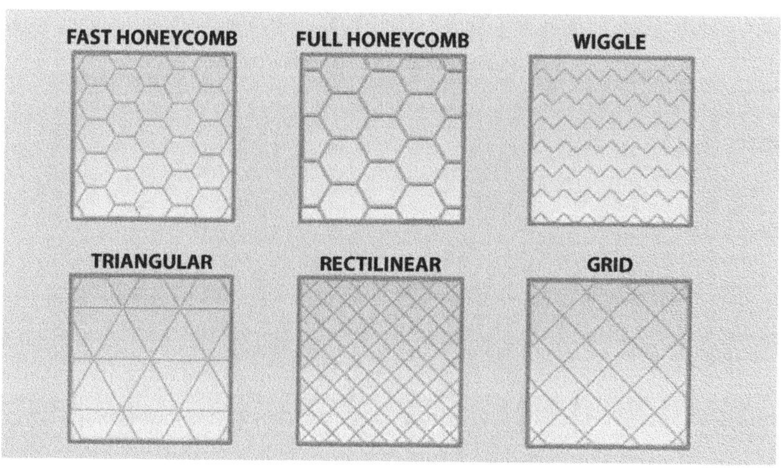


**Figure 14.12** Infill patterns and filament materials.

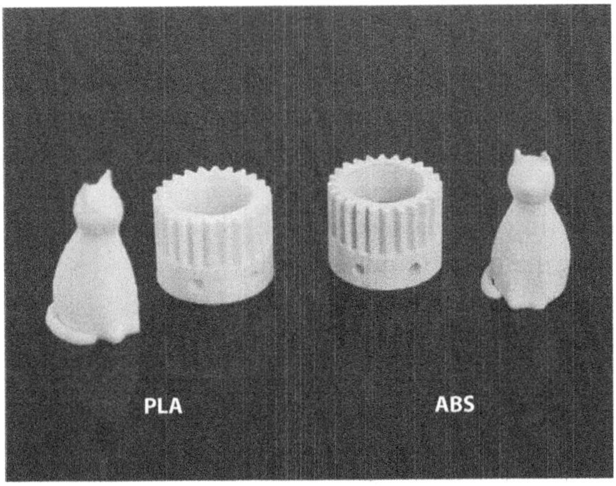


**Figure 14.13** Structure from 3D printing.

### 14.6.5 Basic Information

- The dataset contains 50 rows of data.
- The columns infill_pattern and material consists of categorical entries (infill_pattern = grid or honeycomb and material = abs or pla) instead of numerical entries.
- In the Machine Learning World this is relatively a very small dataset interms of observations. Still we can fit a good regression model out of it and study them. Correlation heatmap of 3D printing Dataset is shown in Figure 14.14.

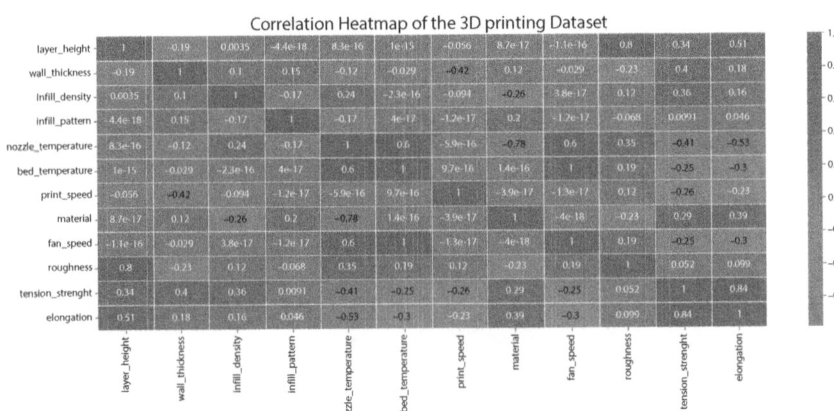


**Figure 14.14** Correlation heatmap of 3D printing dataset.

### 14.6.6 Inference from Heatmap

1. Roughness is correlated to layer height meaning as layer height (feature) increases, roughness increases as well.
2. Material is negatively correlated with nozzle temperarure. So, if nozzle temperature increases, the material quality dimnishes.
3. Bed temperature is correlated with nozzle temperature. As nozzle temperature increases, bed temperature increases.
4. Fan speed is positively correlated with nozzle temperature as well.
5. Fan speed is full positively correlated (1) with bed temperature.
6. The labels elongation and tension strength are positively correlated as well.

linkcode

We will take these heatmap inferences into consideration later when we are in tuning the model. For building the model, we will use all the features in predicting the label. We will use statsmodels library to perform these linear regressions and the result obtain after execution is shown in Figure 14.15.

```
                            OLS Regression Results
==============================================================================
Dep. Variable:              roughness   R-squared:                       0.872
Model:                            OLS   Adj. R-squared:                  0.857
Method:                 Least Squares   F-statistic:                     59.78
Date:                Thu, 07 Jul 2022   Prob (F-statistic):           1.67e-18
Time:                        12:50:37   Log-Likelihood:                -248.88
No. Observations:                  50   AIC:                             509.8
Df Residuals:                      44   BIC:                             521.2
Df Model:                           5
Covariance Type:            nonrobust
=======================================================================================
                          coef    std err          t      P>|t|      [0.025      0.975]
---------------------------------------------------------------------------------------
const                  -0.9307      0.151     -6.172      0.000      -1.235      -0.627
layer_height         1246.5353     83.178     14.986      0.000    1078.901    1414.169
nozzle_temperature     14.7774      2.398      6.163      0.000       9.945      19.610
bed_temperature       -54.3045      8.815     -6.160      0.000     -72.070     -36.539
print_speed             0.5538      0.180      3.070      0.004       0.190       0.917
material              294.1610     56.159      5.238      0.000     180.981     407.341
fan_speed               7.6993      1.177      6.542      0.000       5.328      10.071
==============================================================================
Omnibus:                        0.850   Durbin-Watson:                   1.367
Prob(Omnibus):                  0.654   Jarque-Bera (JB):                0.720
Skew:                           0.285   Prob(JB):                        0.698
Kurtosis:                       2.857   Cond. No.                     3.59e+18
==============================================================================
```

**Figure 14.15** OLS regression result.

[1] Standard Errors assume that the covariance matrix of the errors is correctly specified.
[2] The smallest eigenvalue is 2.48e-31. This might indicate that there are strong multicollinearity problems or that the design matrix is singular.

### 14.6.7 Inference from Model

1. The R-squared value is 0.875 ot 87.5% which states that about 87.5% of proportion of variability in data can be explained by this linear regression model. The r-squared is the most important parameter for a model evaluation. higher the value, better is the fit.
2. The p-value is an important measure for evaluating the various variables. Closer is the p-value to 1, the lesser is that feature variable has to do with the label variable. In our model, wall thickness, infill_density and infill_pattern are all close to 1 meaning they don't play much role in roughness.
3. Finally, the expression for this model can be written from the coefficients as follows

   Roughness = -0.9534 + 1269.4449*layer_height + 2.3342*wall_thickness - 0.0423*infill_density - 0.1255*infill_pattern + 15.0562*nozzle_temperature - 55.6225*bbed_temperature + 0.6496*print_speed + 298.4514*material + 7.8989*fan_speed

### 14.6.8 The Coefficients Suggest the Following

- In the absence of all the parameters, the roughness decreases by 0.9534 micro metre by the constant term. This may be due to other factors like ambient temperature which are not considered for the model.
- A positive coefficient value states that an increase in wall thickness (example) by 1 mm results in increase in roughness by 2.3342 micro metre.
- Similarly, a negative coefficient value states that an increase in bed temperature by a degree C causes 55.6225 micro metre decrease in roughness.

Similarly we will fit models for predicting tension strength and elongation as well. I will reduce the explanations here. The procedure is absolutely same starting with all features and then tuning for better results.

## 14.7 Conclusion and Discussion

There are currently over 100,000 persons on the US national transplant waiting list, and 17 people die every day while waiting for an organ transplant. Bioprinting provides an alternate method for creating tissues/organs for clinical use. Because the bioprinting area is diverse and continuously growing, a working definition should be utilized to characterize it. Three-dimensional bioprinting is an additive manufacturing process that mimics tissue architecture. A typical bioprinting process includes imaging, designing bioprinted tissues, selecting materials (for scaffolding) and cells, bioprinting, and maturing bioprinter tissues. A surgical procedure that included everything from CT scans to 3D printed heart models. However, other obstacles remain, including segmentation, which may be time-consuming, and ensuring the correctness and repeatability of the whole operation. Aside from tissue engineering and regenerative medicine uses, bioprinting has also been utilized to generate 3D tissue/organ models for various biomedical purposes. A bioprinted 3D cancer cell model, for example, has been produced for anti-cancer medication screening. Importantly, a 3D hydrogel-based *in vitro* tumor-stromal model that mimics the angiogenic switch has been developed. This model is particularly effective for investigating the molecular mechanisms behind angiogenesis start. Notable advance in innovative bioink design and bioprinting procedures is necessary in the future to generate biocompatible and completely vascularized tissues/organs. More research is needed, in particular, on the carefully regulated positioning of various types of cells within the scaffold to fully duplicate a natural tissue/organ. The use of artificial intelligence (AI) and machine learning (ML) has enormous promise for enhancing the design of bioprinting processes. Finally, 4D bioprinting is predicted to be the next generation of technology, with a new dimension of change over time (such as functionality altering over time in response to external stimuli). Amid allegations of *in situ* 3D biomaterials, this approach is still in the animal experimentation stage. Furthermore, robots and 3D bioprinting are rarely coupled in the development of novel technology. However, pioneer research in this field has indicated that the integrated technologies in tissue engineering and regenerative medicine have enormous potential.

## 14.8 Future Scope

Nonetheless, there are certain challenges in translating this technology into therapeutic applications. Breakthroughs are required in bioink utilized, robot execution control precision, accuracy, inter harmony, software configuration, device size, and so on. The robotic *in situ* 3D bioprinting technique may take some time to become extensively employed in hospital studies. This chapter is very important for the researchers working in the same field.

## References

1. Whig, P., Velu, A., Naddikatu, R.R., The economic impact of AI-enabled blockchain in 6G-based industry, in: *AI and Blockchain Technology in 6G Wireless Network*, pp. 205–224, Springer, Singapore, 2022.
2. Alkali, Y., Routray, I., Whig, P., Strategy for reliable, efficient and secure IoT using artificial intelligence. *IUP J. Comput. Sci.*, 16, 2, 1–9, 2022.
3. Whig, P., Velu, A., Sharma, P., Demystifying federated learning for blockchain: A case study, in: *Demystifying Federated Learning for Blockchain and Industrial Internet of Things*, pp. 143–165, IGI Global, USA, 2022.
4. Whig, P., Kouser, S., Velu, A., Nadikattu, R.R., Fog-IoT-assisted-based smart agriculture application, in: *Demystifying Federated Learning for Blockchain and Industrial Internet of Things*, pp. 74–93, IGI Global, USA, 2022.
5. Whig, P., Velu, A., Ready, R., Demystifying federated learning in artificial intelligence with human-computer interaction, in: *Demystifying Federated Learning for Blockchain and Industrial Internet of Things*, pp. 94–122, IGI Global, USA, 2022.
6. Whig, P., Velu, A., Nadikattu, R.R., Blockchain platform to resolve security issues in IoT and smart networks, in: *AI-Enabled Agile Internet of Things for Sustainable FinTech Ecosystems*, pp. 46–65, IGI Global, USA, 2022.
7. Whig, P., Velu, A., Bhatia, A.B., Protect nature and reduce the carbon footprint with an application of blockchain for IIoT, in: *Demystifying Federated Learning for Blockchain and Industrial Internet of Things*, pp. 123–142, IGI Global, USA, 2022.
8. Jupalle, H., Kouser, S., Bhatia, A.B., Alam, N., Nadikattu, R.R., Whig, P., Automation of human behaviors and its prediction using machine learning. *Microsyst. Technol.*, 28, 1–9, 2022.
9. Tomar, U., Chakroborty, N., Sharma, H., Whig, P., AI based smart agricuture system. *TLAI*, 2, 2, 25–35, 2021.
10. Whig, P., Nadikattu, R.R., Velu, A., COVID-19 pandemic analysis using application of AI, in: *Healthcare Monitoring and Data Analysis Using IoT: Technologies and Applications*, p. 1, 2022.

11. Anand, M., Velu, A., Whig, P., Prediction of loan behaviour with machine learning models for secure banking. *J. Comput. Sci. Eng.*, 3, 1, 1–13, 2022.
12. Alkali, Y., Routray, I., Whig, P., Study of various methods for reliable, efficient and secured IoT using artificial intelligence, 12, 1–25, 2022. Available at SSRN 4020364.
13. Chopra, G. and Whig, P., A clustering approach based on support vectors. *Int. J. Mach. Learn. Sustain. Dev.*, 4, 1, 21–30, 2022.
14. Chopra, G. and Whig, P., Smart agriculture system using AI. *IJSDCS*, 4, 1, 1–16, 2022.
15. Madhu, M. and Whig, P., A survey of machine learning and its applications. *Int. J. Mach. Learn. Sustain. Dev.*, 4, 1, 11–20, 2022.
16. Mironov, V., Reis, N., Derby, B., Bioprinting: A beginning. *Tissue Eng.*, 12, 4, 631–634, 2006.
17. Archit, A. and Hemant, A., *Yuwei, 3D Printing-Biological Applications*, Presentation, National University of Singapore, Singapore, 2013.
18. Ikada, Y., Challenges in tissue engineering. *J. R. Soc. Interface, The Royal Society, 2006*, Web, 12, 1–19, Sept. 4, 2016.
19. Hölzl, K., Lin, S., Tytgat, L., Van Vlierberghe, S., Gu, L., Ovsianikov, A., Bioink properties before, during and after 3D bioprinting. *Biofabrication*, 8, 3, 1–5, 2016.
20. Sher, D., The top 15 bioprinters. 3dprintingindustry.com, Retrieved 2 May 2018, from https://3dprintingindustry.com/news/top-10-bioprinters-55699/.
21. Melchels, F., Domingos, M., Klein, T., Malda, J., Bartolo, P., Hutmacher, D., Additive manufacturing of tissues and organs. *Prog. Polym. Sci.*, 37, 8, 1079–1104, 2012.
22. Nakamura, M., Iwanaga, S., Henmi, C., Arai, K., Nishiyama, Y., Biomatrices and biomaterials for future developments of bioprinting and biofabrication. *Biofabrication*, 2, 1, 1–15, 2010.
23. Tao, X., Rodriguez-Devora, J., II, Reyna-Sariano, D., Bhuyan, M., Zhu, L., Principles of bioprinting technology, in: *Regenerative Medicine Applications in Organ Transplantation*, 61–67, Elsevier, CA, 2014.
24. Wang, K. and Yuan, Y., Principles of bioprinting technology, in: *Regenerative Medicine Applications in Organ Transplantation*, pp. 67–79, Elsevier, San Diego, CA, 2013.
25. Rauf, S., Kahin, K., Alshehri, S., Abdelrahman, S., Susapto, H.H., Asad, S.J.S., Sunderamoorthy, D., Hauser, C.A.E., *In situ* 3D bioprinting under physiological pH using ultrashort self-assembling peptides, submitted), *J. Mater. Chem.*, 15, 1–16, 2018.
26. Seow, W.Y. and Hauser, C.A., Short to ultrashort peptide nanogels for biomedical uses. *Mater. Today*, 17, 381–388, 2014.
27. Wu, E.C., Zhang, S., Hauser, C.A., Self-assembling peptides as cell-interactive scaffolds. *Adv. Funct. Mater.*, 22, 456–468, 2012.
28. Hauser, C.A.E., Deng, R., Mishra, A., Loo, Y., Khoe, U., Zhuang, F., Cheong, D.W., Accardo, A., Sullivan, M.B., Riekel, C., J.Y.Ying, U.A., Hauser, Natural

tri-to hexapeptides self-assemble in water to amyloid β-type fiber aggregates by unexpected αhelical intermediate structures. *Proc. Natl. Acad. Sci.*, 108, 1361–1366, 2011.
29. Mishra, Y., Deng, R., Chua, Y.J., Hee, H.T., Ying, J.Y., Hauser, C.A., Ultrasmall natural peptides self-assemble tostrong temperature-resistant helical fibers in scaffolds suitable for tissue engineering. *Nano Today*, 6, 232–239, 2011.
30. Ozbolat, I.T. and Yu, Y., Bioprinting toward organ fabrication: Challenges and future trends. *IEEE Trans. Biomed. Eng.*, 60, 3, 691–699, 2013.
31. Chung, J., Naficy, S., Yue, Z., Kapsa, R., Quigley, A., Moulton, S., Bio-ink properties and printability for extrusion printing living cells. *Biomater. Sci.*, 1, 7, 763–773, 2013.
32. Murphy, S. and Atala, A., 3D bioprinting of tissues and organs. *Nat. Biotechnol.*, 32, 8, 773–785, 2014. http://dx.doi.org/10.1038/nbt.2958
33. 3D CAD Design Software. Solidworks.com. Retrieved 2 May 2018, from https://www.solidworks.com/.
34. Wijnen, B., Hunt, E.J., Anzalone, G.C., Pearce, J.M., Open-source syringe pump library. *PloS One*, 9, 9, 1–5, 2014.
35. Organ donation statistics. Available online: https://www.organdonor.gov/learn/organ-donation-statistics (accessed on 12 May 2022).
36. Deo, K.A., Singh, K.A., Peak, C.W., Alge, D.L., Gaharwar, A.K., Bioprinting 101: Design, fabrication, and evaluation of cell-laden 3D bioprinted scaffolds. *Tissue Eng. Part A*, 26, 318–338, 2020.
37. Murphy, S.V. and Atala, A., 3D bioprinting of tissues and organs. *Nat. Biotechnol.*, 32, 773–785, 2014.
38. Bertolini, M., Rossoni, M., Colombo, G., Operative workflow from CT to 3D printing of the heart: Opportunities and challenges. *Bioengineering*, 8, 130, 2021.
39. Benwood, C., Chrenek, J., Kirsch, R.L., Masri, N.Z., Richards, H., Teetzen, K., Willerth, S.M., Natural biomaterials and their use as bioinks for printing tissues. *Bioengineering*, 8, 27, 2021.
40. Lakshmi, T.S., Kasoju, N., Raju, R., Bhatt, A., Formulation and characterization of alginate dialdehyde, gelatin, and platelet-rich plasma-based bioink for bioprinting applications. *Bioengineering*, 7, 108, 2020.
41. Mirzaei, M., Okoro, O.V., Nie, L., Petri, D.F.S., Shavandi, A., Protein-based 3D biofabrication of biomaterials. *Bioengineering*, 8, 48, 2021.
42. Huber, T., Najaf Zadeh, H., Feast, S., Roughan, T., Fee, C., 3D printing of gelled and cross-linked cellulose solutions, an exploration of printing parameters and gel behaviour. *Bioengineering*, 7, 30, 2020.
43. Stenvall, E., Flodberg, G., Pettersson, H., Hellberg, K., Hermansson, L., Wallin, M., Yang, L., Additive manufacturing of prostheses using forest-based composites. *Bioengineering*, 7, 103, 2020.
44. Wang, X., Wang, Q., Xu, C., Nanocellulose-based inks for 3D bioprinting: Key aspects in research development and challenging perspectives in applications-a mini review. *Bioengineering*, 7, 40, 2020.

45. Rosendahl, J., Svanström, A., Berglin, M., Petronis, S., Bogestål, Y., Stenlund, P., Standoft, S., Ståhlberg, A., Landberg, G., Chinga-Carrasco, G. *et al.*, 3D printed nanocellulose scaffolds as a cancer cell culture model system. *Bioengineering*, 8, 97, 2021.
46. Gu, Y., Schwarz, B., Forget, A., Barbero, A., Martin, I., Shastri, V.P., Advanced bioink for 3D bioprinting of complex free-standing structures with high stiffness. *Bioengineering*, 7, 141, 2020.
47. Giubilini, A., Bondioli, F., Messori, M., Nyström, G., Siqueira, G., Advantages of additive manufacturing for biomedical applications of polyhydroxyalkanoates. *Bioengineering*, 8, 29, 2021.
48. Hauser, P.V., Chang, H.M., Nishikawa, M., Kimura, H., Yanagawa, N., Hamon, M., Bioprinting scaffolds for vascular tissues and tissue vascularization. *Bioengineering*, 8, 178, 2021.
49. Jeong, H.J., Nam, H., Jang, J., Lee, S.J., 3D bioprinting strategies for the regeneration of functional tubular tissues and organs. *Bioengineering*, 7, 32, 2020.
50. Nizioł, M., Paleczny, J., Junka, A., Shavandi, A., Dawiec-Liśniewska, A., Podstawczyk, D., 3D printing of thermoresponsive hydrogel laden with an antimicrobial agent towards wound healing applications. *Bioengineering*, 8, 79, 2021.
51. Aleemardani, M., Trikić, M.Z., Green, N.H., Claeyssens, F., The importance of mimicking dermal-epidermal junction for skin tissue engineering: A review. *Bioengineering*, 8, 148, 2021.
52. Kozaniti, F.K., Metsiou, D.N., Manara, A.E., Athanassiou, G., Deligianni, D.D., Recent advancements in 3D printing and bioprinting methods for cardiovascular tissue engineering. *Bioengineering*, 8, 133, 2021.
53. McGivern, S., Boutouil, H., Al-Kharusi, G., Little, S., Dunne, N.J., Levingstone, T.J., Translational application of 3D bioprinting for cartilage tissue engineering. *Bioengineering*, 8, 144, 2021.
54. Naghieh, S., Lindberg, G., Tamaddon, M., Liu, C., Biofabrication strategies for musculoskeletal disorders: Evolution towards clinical applications. *Bioengineering*, 8, 123, 2021.
55. Fuest, M., Yam, G.H., Mehta, J.S., Duarte Campos, D.F., Prospects and challenges of translational corneal bioprinting. *Bioengineering*, 7, 71, 2020.
56. Kuehlbach, C., Hensler, S., Mueller, M.M., Recapitulating the angiogenic switch in a hydrogel-Based 3D *in vitro* tumor-stroma model. *Bioengineering*, 8, 186, 2021.
57. Ghosh, S., Chaudhuri, S., Roy, P., Lahiri, D., 4D printing in biomedical engineering: A state-of-the-art review of technologies, biomaterials, and application. *Regen. Eng. Transl. Med.*, 15, 1–15, 2022.

# 15

# 3D Bioprinting Technology Optimization Using Machine Learning

**Kuldeep Singh Kaswan[1], Jagjit Singh Dhatterwal[2*], Reenu Batra[3], Balamurugan Balusamy[4] and Gangadevi E.[5]**

[1]*School of Computing Science and Engineering, Galgotias University, Greater Noida, India*
[2]*Department of Artificial Intelligence & Data Science, Koneru Lakshmaiah Education Foundation, Vaddeswaram, Andhra Pradesh, India*
[3]*Department of Computer Science and Engineering, SGT University, Gurugram, India*
[4]*Office of Dean Academics, Shiv Nadar University, Delhi, India*
[5]*Department of Computer Science, Loyola College, Chennai, India*

## *Abstract*
Three-dimensional (3D) bioprinting technology has emerged as a revolutionary approach in tissue engineering and regenerative medicine, offering the potential to fabricate complex and functional biological constructs. As the field continues to evolve, researchers have recognized the importance of incorporating machine learning (ML) techniques to optimize and enhance the bioprinting process. This review aims to provide a comprehensive overview of recent trends and advances in machine-learning-optimized 3D bioprinting technology. First, we present an introduction to the fundamental principles of 3D bioprinting, highlighting the significance of bioink formulation, printing techniques, and post-printing processes. Subsequently, we delve into the utilization of machine learning algorithms for various aspects of 3D bioprinting, including image analysis, material characterization, process optimization, and quality control. We discuss the integration of ML algorithms with bioprinting platforms to enable real-time monitoring, feedback control, and adaptive optimization of printing parameters. Moreover, this review explores the emerging applications of machine-learning-optimized 3D bioprinting technology in tissue engineering, organ-on-a-chip systems, drug screening, and personalized medicine. We examine how ML techniques can facilitate the design and fabrication of complex tissue constructs with enhanced

**Corresponding author*: jagjits247@gmail.com

E. Gangadevi, M. Lawanya Shri, Rajesh Kumar Dhanaraj and Balamurugan Balusamy (eds.) Computational Intelligence in Bioprinting, (303–322) © 2024 Scrivener Publishing LLC

biological functionality, structural integrity, and biocompatibility. Additionally, we discuss the challenges and future prospects of integrating machine learning into the bioprinting workflow, including data acquisition, algorithm development, and regulatory considerations. Overall, this review highlights the transformative potential of machine-learning-optimized 3D bioprinting technology in advancing the field of tissue engineering and regenerative medicine. By leveraging the capabilities of machine learning, researchers can harness the power of data-driven optimization, enabling faster and more accurate bioprinting processes. With further advancements in ML algorithms, data acquisition techniques, and interdisciplinary collaborations, the integration of machine learning and 3D bioprinting holds promise for the development of next-generation biofabrication systems with unprecedented capabilities.

***Keywords:*** Machine learning, 3D bioprinting, computer aided design

## 15.1 Introduction

3D bioprinting, an innovative kind of additive manufacturing inspired by tissue engineering, prints 3D items from biodegradable polymers including living cells (i.e. bioinks) [1]. In 3D bioprinting, biological materials or bioink are extruded instead of molten plastic, which is then cooled and solidified. Because hydrogels create a highly moist environment that fosters cell growth, they are the most widely used bioink component. Most matrix bioinks are natural-origin hydrogels made from substances like collagen, hyaluronic acid, or alginate. Bioink often makes use of materials based on gelatin methacrylate, collagen, poly(ethylene glycol), pluronic, alginate, and decellularized extracellular matrix [2]. As 3D bioprinting permits the production of intricate structures that cannot be constructed using standard manufacturing procedures, it offers hope for resolving the organ shortage problem in transplanting. The bioprinted scaffolds might be turned into functional tissues or organs by activating biological processes using a variety of growth agents for restoration. In addition, doctors may find the bioprinted tissues or organs helpful while training for intricate surgical operations in a simulated setting. Due to its capacity to recreate a broad spectrum of complex human tissues, 3D bioprinting technology is gaining popularity among surgeons and medical device makers for teaching and testing purposes. By allowing for the creation of more realistic surgical simulation models, the bioprinter will aid in the improvement of surgical skills and the reduction of surgical mistakes [3].

There are several 3D bioprinting techniques available today. The extrusion technique is the most often used for bioprinting. By applying the right amount of force to the biomaterials, scaffolds may be created without the use of harmful organic solvents. Inkjet bioprinting has the potential to print a wide variety of cells in a controlled manner. Stereolithography bioprinting makes use of a photocurable liquid resin, which is crosslinked using UV light. There are three primary stages in the bioprinting process: pre-bioprinting, during-bioprinting, and post-bioprinting. At the stage prior to printing, computer-aided design (CAD) software of many kinds is employed [4]. After the CAD design is finished, it will be "sliced" into an STL file, which can then be used to generate G-code for manufacture. G-code is a computer language used to provide instructions to printers connected to computers. The most efficient design is one that ensures maximum nutrient and oxygen delivery to as many tissues or organs as possible. The comparison between the original concept and the data is normally distributed would provide a plethora of information. The biomaterials and bioprinters used to make the final product have an effect on its rheology, biology, and degradation. Biocompatibility is a significant consideration when selecting a biomaterial due to the fact that biomaterials interact with living tissues in the body or bodily fluids, and the production of a toxic or immunological response might be lethal. Researchers would investigate real-time techniques of defect detection in printed artwork. Following printing, research into storage and continuous quality monitoring procedures ensures that cells continue to grow in ideal settings. This is a major challenge for 3D bioprinting and general 3D printing since it makes it hard to determine the best printing parameters [5].

As a solution, machine learning (ML) techniques were first created for the 3D printing sector. Machine learning (ML) is a branch of artificial intelligence that aims to imbue machines with cognitive experiences (e.g., the capacity to learn). Algorithms designed for machine learning (ML) might potentially self-train by analyzing data and adjusting their predictions and overall performance accordingly. These algorithms don't carry out a predefined set of procedures but rather build an ideal model from which inferences and decisions [6]. By combining this advanced computer technology with cutting-edge medical equipment, synergy may be established. In certain cases, the ML-based model may achieve better performance than the physics-based one. The conventional physics-based model has issues due to its non-linearities, varying parameters, and unknowns. Nonetheless, ML excels at prediction, has versatile techniques, and is very adaptable. The choice between an ML model driven by data and one based on physics is a vital one. The answer will depend on the specifics of the problems at hand. Without first-hand data, mathematical models of a system's operation are useless for making reliable predictions about that system's behavior. Nevertheless, if there are numerous example outcomes, the ML-based model may be employed with training, validation, and testing datasets [7]. In machine learning, the cost function is used to measure a model's performance in comparison to the true values. The loss function is defined by a single data point, whereas the cost function is defined by a complete data instance. The optimal values for the model's parameters are determined by the optimization algorithms in the cost function, which search for the minima of the cost functions. Due to the nature of optimization, a wide range of printing factors may be tweaked to obtain the best possible design and printing environment within a given set of limitations [8].

Several people are interested in employing ML for additive manufacturing (3D printing) since it allows for the production of components with granular printing controls. Many areas of 3D printing were improved by using ML technology; they included filament size, melting/bed temperature, printing speed, layer thickness, infill density, and nozzle size. Despite the many potential applications, there has been very little study of ML's role in 3D bioprinting processes in comparison to more conventional forms of 3D printing. Recent advances in biomedical engineering have seen the emergence of 3D bioprinting as a promising medical application, which may be used in tandem with other promising new technologies such as telemedicine, the healthcare metaverse, and medical wearables. The improved print quality made achievable by ML might one day be used to repair or replace damaged organs, provide more accurate drug testing, and lessen the need on animal testing. The expanding field of machine learning–based bioprinting holds great promise for regenerative medicine. This paper provides a summary of studies conducted between 2016 and 2021 on the application of ML to 3D bioprinting, as well as studies conducted between 2015 and 2021 on the usage of 3D bioprinting systems. This might provide researchers with inspiration for new ways to use ML to advance 3D bioprinting [9].

## 15.2 Human Organs Printed Through 3D Printers

3D printing has a significant influence on many fields and industries, including production, medical, robotics, transportation, automotive, aerospace, and education. Human tissues and organs may be printed using 3D bioprinters using bioinks composed of nanomaterials and cells for tissue engineering. Microscale tissues, also known as organoids, are created to study diseases and identify new drugs and they may be utilized to create functional substitutes for injured tissues [10].

### 15.2.1 Prioritize Cell Membrane

There is no way around using bioinks in bioprinting. Bioinks are used to fabricate tissue scaffolds and are mostly composed of cells and hydrogel pre-polymers. It's important to prioritize the cell membrane. The methodologies and applications of printing dictate that the cells used for bioink should mimic the physiological state *in vivo*. After then, the cells may be housed by the biomaterials while they continue to perform their original functions. Researchers employ stem cells, primary cell lines, and other comparable cell sources derived from humans and other animals [11]. Enzymatic or mechanical methods are used to separate primary cells from animal or human tissue. Stem cells are superior to other cell sources for bioprinting because of their greater proliferative capacity and ability to appropriately mimic the function of the original cells. Most of these cells can be obtained with relative ease, however due to the limitations of current bioprinting processes, some may prove more challenging to cultivate. Thus, bioprinting labs focus on developing cutting-edge cell technologies for use in research and industry [12].

The biomaterial's viscosity and rheological behavior impact the quality and shape of the printed structure. Thus, it is essential to choose appropriate biomaterials prior to printing. Hydrogels have gained popularity as a medium to mold cells for 3D bioprinting due to their high biomedical applications and ease of processing. Hydrogels are crucial in bioprinting. Gels that are aqueous are called hydrogels, a combination of the terms hydro (water) and gel. They are impermeable polymer frameworks that can absorb water without changing their shape and expand to an equilibrium volume [13]. Hydrogels thereby dominate the biomechanical and biodegradable properties of the manufactured object and come into intimate contact with cells, functioning as an extracellular matrix-like microenvironment. As cells divide and multiply, new tissue forms, and the extracellular matrix plays a key role in this process. It is recommended that a hydrogel pre-polymer solution used in bioprinting be printable and crosslinkable. As seen in Figure 15.2, pre-polymers are commonly required to have shear thinning properties or carbon-double bond topologies. It is just as crucial that the pre-polymer solution be biocompatible as it is that it be printable. Crosslinking may be achieved physically, chemically, or enzymatically during the hydrogelization of a pre-polymer solution. Physical crosslinking takes use of non-covalent connections such hydrophobic interactions, coordination bonding, ionic contact, and van der Waals attraction, as opposed to chemical crosslinking, which relies on covalent connections [14]. Thermosetting polymers result from chemical crosslinking, whereas thermoplastic elastomers are an example of physical crosslinking. To chemically crosslink is to form a covalent connection between two or more molecules, either intermolecularly or intramolecularly. When polymer chains are crosslinked, they form a stronger network via a combination of noncovalent and covalent interactions. As shown in Figure 15.1, many bioprinters are recommended for use with different crosslinking techniques.

In bioprinting, hydrogels are often made from pre-polymers, such as biopolymers or synthetic polymers. Biopolymers are biodegradable materials that are synthesized by cells in a living organism. Cellulose, alginate, collagen, and chitosan are examples. These substances originate from either plants or animals [15]. Extrusion printing works well with many natural biopolymers because of their non-covalent interactions. This results in a fluid with the rheological properties of shearing luminosity and thixotropy, both of which are desirable. The biopolymers' capacity to spontaneously coat eukaryotic cell surfaces by combining with proteins is a benefit. They are biodegradable and responsive because they are able to replicate the organic matrix of normal tissues [16].

On the other hand, synthetic fibers are a product of scientific and technological innovation, made from petroleum oil. Physical and chemical features, such as crosslinking rates and mechanical properties, may be tailored to a specific application. Common materials used in 3D printing include polyethene glycol, polyacrylamide, and poly (N-isopropyl acrylamide). The polymer or monomer-containing ink used for hydrogel printing ideally has rheological or light-curing characteristics. Hydrogels created by printing should be safe for cells and easy to maintain. In addition, hydrogels must have exceptional mechanical properties. Combining some biomaterials with cellulose, polyvinyl alcohol, cellulose fibrils, and chitosan increases their viscosity and shear-shining characteristics through molecular interactions. The light-curing capability of biomaterials may be fine-tuned in other ways as well Methacrylates, such as gelatin methacrylate and hyaluronic acid methacrylate, may be used to modify biopolymers by grafting carbon-carbon double bonds onto the polymer chains. Because to the existence of carbon-carbon double bonds, these modified polymers are more suited to photopolymerization print modes than native biopolymers.

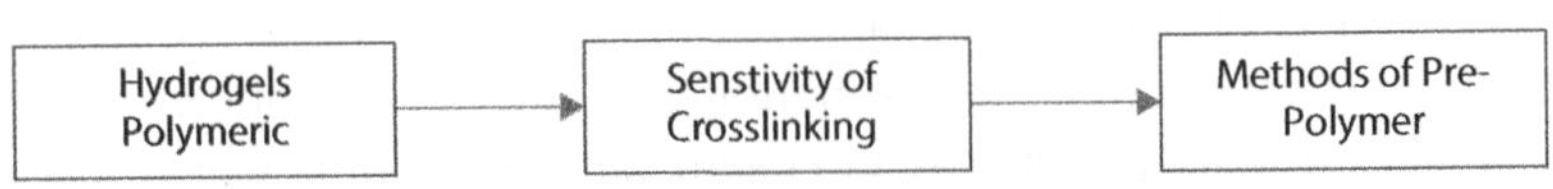


**Figure 15.1** Structure of 3D bioprinting.

These polymers may be crosslinked with either ultraviolet (UV) or visible (visible) light, combining the best features of both biopolymers and synthetic polymers. Despite the rapid development of printing technology, these materials are not yet sufficient to build complex human tissue/organs. Finding an appropriate hydrogel for 3D bioprinting is an intriguing and productive endeavor [17].

Modalities: Here, we provide a categorization and analysis of four prominent 3D bioprinters. Bioprinting techniques range from inkjet printing to laser-induced forward transfer (LIFT) printing to scaffold-based printing to extrusion-based printing. Extrusion-based bioprinting is a simple 3D bioprinting technology because bioinks are manufactured under shear stress to deposit 3D objects. Extrusion-based bioprinting often employs polymers like alginate or fibrin as bioinks. They can be easily manipulated after being sheared and extruded via a small nozzle. In addition, they have built-in cellular adhesion molecules, which make it easier to link cells physically and guarantee the structure's stiffness once printed [18].

A cells are embedded in bioink and put into the syringe. Shear stress is imparted to the bioinks through nozzles, allowing them to be extruded into 3D shapes. As bioink is extruded, it adopts the exact shape that was envisioned in a CAD program. There are many different approaches to 3D printing, but most of them need the use of computer-aided design (CAD). As CAD (computer-aided design) allows for the development of digital models of actual items, it is an essential element of 3D printing. Separate layers may be extracted from an STL-formatted 3D model. After that, the application will generate the tool path (G-code). This is a vital process because it converts the 3D model into a form that the 3D printer can read and reproduce. This file instructs the 3D printer on how much and where to deposit the material of choice [19].

Scaffold-free bioprinting has gained a lot of attention as a viable alternative to the scaffold-based technique. Repairs of bone, cartilage, and ligaments have been performed using scaffold-free procedures in clinical settings. One benefit of the scaffold-free method is that cells naturally build a strong intercellular connection. This method might be used to print complex organs and tissues with a high cell concentration [20]. Scaffold-free bioprinters produce cell aggregates or rods without any further support structures. These clumps are able to expand into more substantial structures because to the fluidity and fusion of tissues throughout time. For cells to self-assemble into progressively large structures, interactions between cells and molecules are essential [21].

Inkjet printing is inexpensive because commercial printers (i.e. office inkjet printers) may be transformed into inkjet bioprinters. 3D inkjet bioprinting uses piezoelectricity or heat to apply the ink, much as traditional inkjet printing. Biological structures may be printed using 3D inkjet printers by depositing bioinks layer by layer in a specific pattern, as shown in Figure 15.3c. Investigation into the mingling of individual ink droplets is the most crucial part of 3D inkjet bioprinting. The tiny size of printable droplets poses challenges for the bioprinting of large structures like organs. The sharpness of printed patterns, however, benefits greatly from the tiny droplet size [22].

LIFT bioprinting (Figure 15.3d) is similar to inkjet bioprinting in that bioinks are projected from a donor thin film of biopolymer towards a receiving platform, but instead of an inkjet nozzle, a laser beam is used as the driving mechanism. Nozzle clogging is perhaps less of an issue with the LIFT bioprinter since it does not utilize nozzles. The LIFT printer has the ability to print a more exact structure at a faster rate, but it requires more money to set up and is not fully automated. There are advantages and disadvantages to all four of these bioprinters. While determining the best printing method, it is important to take into account the characteristics of the tissue/organ, cell, and biomaterials being printed [23].

### 15 2.2 Prediction-Based Algorithms

The term "machine learning" (ML) has been in use and refers to research into computer systems that may potentially improve themselves via experience and data. Machine learning (ML) is a subfield of AI that allows computers to draw conclusions and make decisions without being given a set of rules to follow. You may classify ML as either supervised, unsupervised, semi-supervised, or reinforced based on the kind of signal or feedback available to the learning system. In the next paragraph, describe the characteristics of each ML variety [24]. Using supervised ML, we developed classification algorithms that can distinguish between healthy and sick leaves in the field with a success rate. They also employed deep learning (DL) techniques (such as AlexNet, SqueezeNet, GoogLeNet, etc.) to classify the leaves as healthy or sick with more accuracy than ML. Advanced multi-layered neural networks are used in DL, which is a subset of ML. Communication occurs between the various tiers through weighted channels. The dataset is typically split into a training set and testing set at a fixed ratio. For very big datasets, 80/20 proportions between training and testing are recommended. Nevertheless, for more manageable samples, splits are recommended. Researchers advise employing at least many instances, whereas the bare minimum is 1000. Many people are interested in ML and DL because they have the potential to save labor while enhancing the precision of predictions and judgments. In this article, we'll examine the four main types of ML and DL and go through the research that has used ML to 3D printing and 3D bioprinting [25]. The most widely used method of machine learning (ML) needs labeled datasets in order to provide accurate outputs for mapped inputs. Supervised machine learning has many practical uses, such as spam filtering in email inboxes, public-space face recognition, and AI-assisted medical diagnosis. Supervised ML entails an algorithm being fed a training dataset and then using the validation dataset to draw inferences about the accuracy of the training dataset's labeling predictions. Data for supervised ML may be obtained from preexisting instances or generated explicitly. In addition to aiding in problem-solving, the collected knowledge may be put to use to raise efficiency benchmarks. The subfields of classification and regression may be drawn out of supervised ML depending on whether the labels are discrete or continuous [26]. Classification yields results that are expected by the categories or labels applied to them, whereas regression often yields findings that are predicted by the characteristics of the data. Binary classification may have two categories, while multi-category classification can have many (multi-classes). Supervised ML makes use of several algorithms such as naive bayes, decision tables, support vector machines, random forest, k-nearest neighbor, etc. The data collection process is the most time-consuming part, especially when dealing with massive amounts of information. Significant challenges also exist during the preparation of data. Notwithstanding these limitations, supervised ML may be controlled with high precision [27]. In tabulate several supervised ML techniques used in 3D printing. Additive manufacturing, which includes bioprinting, the lens of studies on supervised ML algorithms. When a dataset lacks labels or a clear structure, researchers turn to unsupervised machine learning (ML). As compared to supervised ML, unsupervised ML is better able to make use of massive datasets. Unsupervised learning algorithms, on the other hand, must find patterns in an unlabeled dataset by themselves. Unsupervised ML may be used to find representative features before using supervised ML with powerful features. Unsupervised ML creates muddled input dataset structures to learn more about the data. In a business setting, such unsupervised ML characteristics might be used to group consumers who have similar tastes, in the natural sciences to categorize closely related species of plants and animals, and in the cyber security industry to spot anomalous network activity. Unsupervised ML includes methods like clustering, anomaly detection, associations, and autoencoders. k-means clustering, self-organizing maps, and restricted Boltzmann machines are only few of the unsupervised ML methods used in 3D printing that are described in a review article on ML in additive manufacturing [28].

Semi-supervised machine learning is a hybrid approach that draws from both supervised and unsupervised ML. During training, semi-supervised ML utilizes both a limited amount of labeled data and many unlabeled data. The performance of a classification system may be improved by using unlabeled data with supervised learning data. To better decide which cluster new data should be added to, labeled data may be helpful for clustering. Semi-supervised ML is used to improve classifier performance when the quantity of labelled data is limited by including unlabeled data. As ML has developed, labeling has gone from being a simple, repetitive task to one that now needs a great deal of knowledge. The popularity of semi-supervised ML may be attributed to its efficiency while dealing with partially labeled data [29].

This semi-supervised ML relies on a number of assumptions to improve performance. The first premise is that everything is perfectly smooth. Labels y1 and y2 are assumed to be equivalent if and only if inputs x1 and x2 share a high probability density region. It is assumed in the low-density setting that the decision boundary of the model does not go across areas of high probability density. The last section introduces the assumptions, which center on the low-dimensional placement of high-dimensional input data and a unique structure known as a manifold. In brief, the smoothness assumption holds if the data values in a cluster are very close to the input space. Let's imagine that an area of high probability is modeled as a collection of points. The low-density assumption is then converted into a manifold assumption by inspecting the data points on the low-dimensional manifold as a cluster. These rules and the assumptions make it possible to make use of unlabeled data. Well-liked approaches include, but are not limited to, self-training algorithms, minimum-margin algorithms, and perturbation-based algorithms. Self-training uses a simple pseudo-labelling approach and is the simplest kind of semi-supervised ML. Only labelled data are used in the first stages of training, whereas pseudo-labeled copies of unlabeled data are used in subsequent stages [30].

The purpose of a reinforcement learning system is to teach a target behavior by rewarding or punishing it. Instead of using training data to improve performance as supervised/unsupervised ML does, state-aware ML looks at the outcomes of actions taken in that state. As a result, reinforcement ML involves repeating training operations in order to get better rewards. The environment, or numerical simulations thereof, must be suitable, and sufficient computer resources must be available, as must well specified rules for praise and encouragement and negative reinforcement.

Reinforcement ML employs both supervised and unsupervised ML methods to establish causal relationships between inputs and outcomes [31].

Previously, we discussed how Reinforcement ML may be implemented as a Markov decision process that sequentially performs actions inside a predetermined environment. It has little in do with traditional supervised ML. The focus is on online effectiveness rather than the explicit correction of improper behavior, and prior knowledge of the Markov decision process is not required. In other words, reinforcement ML strikes a balance between relying on established norms and venturing blindly into the unknown. Reward and punishment are automatically determined based on the current state and behavior. This kind of education is influenced by behavior modification and may be used for a variety of purposes, including but not limited to operational science, computational complexity, and simulation-based optimization. As DL is a subset of ML, self-learning techniques are used for both DL and ML. Nowadays, the term has been used to a broader range of machine learning methods outside DL. To fully grasp DL, it is necessary to first familiarize oneself with ML's methodology. The machine learning procedure consists of four stages: data gathering, variable processing, model training, and model assessment. Even without the variable processing step, DL is projected to provide respectable results in this scenario. Yet, this does not imply that DL is always effective when ML is ineffective. The task at hand, the nature of the problem at hand, and the quantity of data available all play a role in deciding which one to use. Although DL is effective when working with large datasets, it does so at the expense of a high-powered computer and some time. Data augmentation is a helpful technique for managing enormous data sets. It provided the bioprinter with DL algorithm parameters intended to mitigate the impacts of cell-induced light scattering. Light dispersion from cells in the bioink has been a persistent problem for bioprinting methods that rely on digital light processing, and their work has helped lessen this problem. DL methods perform well when neither extensive domain knowledge nor a thorough understanding of the variables are available. In addition, DL is very effective at addressing challenging problems like image classification, natural language processing, and speech recognition.

### 15.2.3 Bioimaging through CAD Software

In order to properly bioprint a tissue or organ, there are normally three steps involved: pre-printing, printing, and post-printing. In a nutshell, how an organ is printed. Bioimaging and CAD software were used to generate an STL file, which was then cut into binary slices for the 3D printer. In order to print a body part or tissue, one must first prepare the bioinks, then design a slice of one layer, and then print all of the layers. The bioink's conceptual layout is shown in Figure 15.2. PBS may be used to remove any remaining ink after printing. The bioink's cell would grow and divide in a bioreactor to mimic the organ's functions. While 3D bioprinting has come a long way, it is still challenging to produce a completely functional, large-scale organ. So, it is possible to employ ML technology in conjunction with these three methods to improve the quality of complex printing.

Conceptualization of Bioink (adapted from with permission).

Pre-printing: Pre-bioprinting preparation is crucial due to the nature of the bioprinting structure. Planning ahead and figuring out the right conditions are essential for producing anatomically correct tissue models and ensuring the cells are of high quality. There has been a significant uptick in the use of ML methods to the field of 3D bioprinting. In this piece, we'll explore the potential applications of machine learning algorithms in the groundwork for bioprinting.

Bioprinting requires a bioink that is both printable and biologically compatible; inductive logic programming (ILP) and multiple regression were utilized to achieve this goal. Multiple regression was used to foretell bioink's printing capacity. Bioinks were made by combining hyaluronic acid and fibrin with three types of natural collagen and atelocollagen. The printability, shape retention, nozzle clogging, frequency dependent storage (G0), and shear stress of the bioink were all examined. Hence, larger G0 is necessary for stronger form integrity, while lower shear stress is needed for extrusion. This study should provide the groundwork for further exploration and development of bioink in the future. The design parameters for piezoelectric drop-on-demand were optimized using ML, which was fed a multi-objective optimization problem. There are a few problems with this printing method: it's not very precise or stable, the minimum print height is rather high, and the print speed suffers because of the large droplet size. We used fully connected neural networks to establish a relationship between the various printing parameters. They also looked at the hybrid multi-sub gradient descent bundle method, namely its adaptive learning

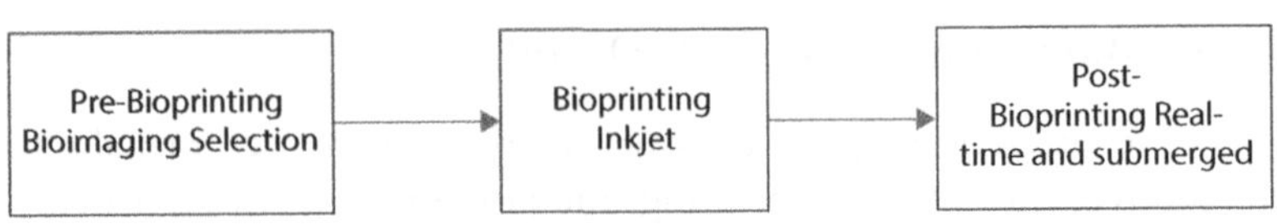


**Figure 15.2** The schematic of developing the bioink.

rate algorithm. As a result, they achieved consistent printing of a single droplet at a print speed that increased from 0.88 to 2.08 m/s1. This finding offers the possibility for the precise execution of cell structures and complex biological activities on bioprinting platforms now under development.

The Proper Way to Print: It's possible that ML algorithms may help with a number of important printing parameters. Process and component quality optimization, as well as anomaly prediction, are two examples of solvable issues. As bioink quality varies widely across different bioprinters, it's important to choose a bioprinter well-suited to the needs of the research project and the surrounding environment before commencing the printing process.

The four primary bioprinting modalities are extrusion-based bioprinting, scaffold-free bioprinting, inkjet bioprinting, and LIFT bioprinting. When it comes to 3D bioprinting, ML can provide guidance on anything from which bioinks to use to which 3D bioprinters and software to use. Complex procedures may be taught using ML techniques. As a result, ML will improve production by fine-tuning existing settings and looking ahead to possible outcomes. It developed a convolutional neural network-based anomaly detection method to classify defects in hydrogel-based bioink. Grids, spheres, lines, and honeycombs were all made by processing images into smaller and smaller parts. High-quality tissue composition is anticipated to be printed in 3D using this method, with real-time autonomous correction.

Significant parameters in extrusion bioprinting include the printed filament diameter (d), nozzle diameter (D), nozzle length (L), apparent viscosity of bioinks (m), stage moving speed (v), and shear stress (S). It is possible to optimize bioprinting in terms of both print quality and speed by modelling the underlying relationships. Many research have looked at the links between these factors. You need to know the nozzle diameter (D), apparent viscosity of bioinks (m), nozzle length (L), and stage movement speed (v) to calculate the expected size of your manufactured filament (v). Factors such as printed filament diameter and the implantation of appropriate cells in the scaffold influence the resolution of a bioprint. However, ML and DL demonstrated that there was a lack of studies probing the factors linked to cell viability and proliferation. Figure 15.3 shows how a neural network is trained with the use of printing parameters. Algorithm settings are then adjusted in light of useful results (such as cell viability and cell growth). The topology of a neural network permits reliable predictions of outputs given just a set of inputs that themselves are unknown.

Cell viability is negatively impacted by laser time and intensity during LIFT bioprinting, yet these factors are necessary for crosslinking and

**Figure 15.3** Examples of consideration in bioprinting for a neural network optimization.

mechanical features of the biomaterial in the case of hydrogels. By training a neural network using parameter (laser)-outcome (cell and mechanical properties) data, we may maximize the degree of crosslinking and cell survival. Based on this trend, it is likely that ML will play a significant role in the development of 3D bioprinting [32].

Several types of manufacturing errors are simple prey for a convolutional neural network. Bioprint quality might be improved using ML and real-time monitoring. Approximate dimensions and curved layers may be identified, and mistakes like improperly positioned bioinks or damaged microstructures can be repaired. As edges and lines are independent of an object's location, they are ideal candidates for extraction by convolutional neural networks. Error detection and correction in 3D bioprinting via the use of neural networks with convolutional structure. High-quality input images were used to train the convolutional layers. The data was analyzed by the model, and labels were applied based on its conclusions. Bioprinting's faults were finally revealed via classification techniques.

The time immediately after printing is crucial for evaluating the quality of bioprinting technologies' final products. There is sometimes a discrepancy between the planned printed output and the actual outputs, especially in regards to design and cell development. Even if the plan was designed and assembled correctly before printing, the printed materials may never reach the patient or surgeon in time for usage.

Up to eight weeks of rigorous tissue maturation monitoring may be required for tissues/organs made utilizing 3D printing technology to reach full functionality. The circumstances under which cells are stored (pH, temperature, gas concentration, nutrition supply, waste disposal, and delivery of physical stimulation) are crucial to their continued viability. Experimental evaluation of potential maturity determinants is necessary for promoting tissue maturation and maintaining stable tissue features. Hence, the cells, bioprinters, and printing quality should be considered as much as possible during post-printing processing. When these conditions are satisfied, ML becomes the most innovative approach.

When cells, growth regulators, and biomaterials are combined on a scaffold, a microenvironment is created in which tissue architectures may form. Scaffolds serve several purposes, including as cell adhesion, nutrient transport, and active cell-biomaterial interaction, and are therefore notoriously complex. With ML, we may reliably estimate the success rate of scaffolds in construction, enabling for the building of high-quality scaffolds. As a result of ML's quantitative evaluation of the printability of producing a scaffold matched to the individual's skeletal structure, the advantages for various printers are enhanced. An enhanced printer setup using a probabilistic model was developed, complete with a grading system for filament. Training data included variations in printing conditions (extrusion pressure, bioink temperature, printing speed) and bioink concentrations. Bioinks containing gelatin methacrylate and hyaluronic acid methacrylate were evaluated for their printability using Bayesian optimization. A pore structure was designed on the layer stacking of filament morphology, and a scoring system was implemented during extrusion to objectively evaluate printability.

## 15.3 Predictive Trial and Error 3D Printing

The goal of machine learning (ML), a branch of AI, is to use previously discovered patterns in data for predictive or classification purposes. In 3D bioprinting, ML is a potent tool due to its capacity to improve print quality and detect anomalies. This advancement considerably reduces the time and bioink that would have been wasted on trial and error printing. By designing a suitable environment for cell development, scientists may simulate the structure of human tissues and organs.

Nevertheless, there aren't as many studies based on bioprinting as there are on regular printing with ML. It is challenging to collect significant amounts of data because of the biomedical nature of the area. ML requires a large amount of data for training and testing in order to achieve high classification accuracy. Yet, as open source develops, more data will be made accessible to the public. As researchers start storing and sharing their data on the cloud, it is expected to greatly simplify the data collection process. For instance, CAM software may be used to create and use unique 3D models for usage in a digital context by incorporating data from diagnostic imaging procedures like computed tomography and magnetic resonance imaging. If patients desire the greatest potential results from their medical treatment, they should reject a uniform, standardized therapy in favor of personalised care.

The scalability of AI is another unanswered question. As trained ML models are so dependent on the mathematical features of the training data, it is not guaranteed that they will always perform similarly in terms of their abilities. To increase the applicability of AI, it is necessary to better choose data via the use of data scaling techniques like normalization and standardization. Also, because to shared characteristics, transfer learning facilitates AI's rapid adaptation to new domains. Several significant worries, meanwhile, center on questions of information quality and model generalizability of the findings. We require state-of-the-art ML, a hybrid classical mathematical model, and a digital twin to solve these problems and develop trustworthy 3D bioprinting methods. The digital twin is a simulation that may be used to perform tests in place of actual hardware. The term "in silico experiment" refers to a study that takes place purely in a digital environment. Bioprinting and ML's potential together were discussed. They concentrated in on closing the gap between big data and a digital twin, which resulted in in silico and digital trials in the bioprinting space. Collecting publicly available massive data pertinent to bioprinting, or substantial data curation, is the first step toward overcoming the complexities of bioprinting and leading to in silico experimentation. Creating digital twins of tissues/organs requires a deep understanding of physiological health and replication for cellular resolution and cell properties. The exceptional quality of the bioink suggests that one day 3D bioprinting may be done totally digitally. The bioprinting industry as a whole might benefit from digitalization and automation brought about by big data and digital twins since they would make bioprinting simulations possible. Numerous industries have begun using AI, and ML is proving it can capitalize on this growing trend. The continuous improvement of current technology is a driving force behind the future of 3D bioprinting.

## 15.4 Conclusions

In conclusion, the integration of machine learning (ML) techniques with three-dimensional (3D) bioprinting technology has the potential to revolutionize the field of tissue engineering and regenerative medicine. This review has provided an in-depth analysis of recent trends and advances in machine-learning-optimized 3D bioprinting, highlighting its significance and potential applications. By leveraging ML algorithms, researchers can enhance various aspects of the bioprinting process, including image analysis, material characterization, process optimization, and quality control. The real-time monitoring, feedback control, and adaptive optimization enabled by ML algorithms result in improved accuracy, precision, and efficiency in bioprinting processes. This advancement paves the way for the fabrication of complex tissue constructs with enhanced biological functionality, structural integrity, and biocompatibility. Moreover, the integration of machine learning into 3D bioprinting opens up new possibilities in tissue engineering, organ-on-a-chip systems, drug screening, and personalized medicine. ML-optimized bioprinting technology enables the creation of more physiologically relevant models for studying disease mechanisms, drug responses, and personalized therapies. This has the potential to revolutionize drug development, reduce animal testing, and improve patient-specific treatments. However, there are challenges that need to be addressed for the widespread adoption of machine-learning-optimized 3D bioprinting. These challenges include acquiring large-scale, high-quality data for algorithm training, developing robust and interpretable ML models, and addressing regulatory and ethical considerations. Collaboration between experts in bioprinting, machine learning, and regulatory bodies is crucial for overcoming these challenges and ensuring the safe and effective implementation of ML in bioprinting workflows.

## References


1. Zhang, Y.S. and Khademhosseini, A., Advances in engineering hydrogels. *Science*, 356, 6337, eaaf3627, 2017.
2. Lee, V.K., Kim, D.Y., Ngo, H., Lee, Y., Seo, L., Yoo, S.S., Vincent, P.A., Creating perfused functional vascular channels using 3D bio-printing technology. *Biomaterials*, 35, 28, 8092–8102, 2014.
3. Zhang, Y.S., Arneri, A., Bersini, S., Shin, S.R., Zhu, K., Goli-Malekabadi, Z., Khademhosseini, A., Bioprinting 3D microfibrous scaffolds for engineering endothelialized myocardium and heart-on-a-chip. *Biomaterials*, 110, 45–59, 2016.
4. Kaswan, K.S., Dhatterwal, J.S., Kumar, K., Blockchain of internet of things-based earthquake alarming system in smart cities, in: *Integration and Implementation of the Internet of Things Through Cloud Computing*, pp. 272–287, IGI Global, Elsevier, 2021.
5. Murphy, S.V. and Atala, A., 3D bioprinting of tissues and organs. *Nat. Biotechnol.*, 32, 8, 773–785, 2014.
6. Ng, W.L., Lee, J.M., Zhou, M., Chen, Y., Xu, K., Wang, Y., Yeong, W.Y., Volumetric bioprinting of complex living-tissue constructs within seconds. *Adv. Mater.*, 32, 43, 2003991, 2020.
7. Rivenson, Y., Wang, H., Wei, Z., de Haan, K., Zhang, Y., Wu, Y., Ozcan, A., Virtual histological staining of unsectioned tissue-autofluorescence images via deep learning. *Nat. Biomed. Eng.*, 3, 6, 466–477, 2019.

8. Koch, L., Deiwick, A., Schlie, S., Michael, S., Gruene, M., Coger, V., Chichkov, B., Skin tissue generation by laser cell printing. *Biotechnol. Bioeng.*, 109, 7, 1855–1863, 2012.
9. Hoch, E., Tovar, G.E., Borchers, K., Bioprinting of artificial blood vessels: Current approaches towards a demanding goal. *Eur. J. Cardiothorac. Surg.*, 45, 5, 753–762, 2013.
10. Nair, K., Gandhi, M., Khalil, S., Yan, K.C., Marcolongo, M., Barbee, K., Sun, W., Characterization of cell viability during bioprinting processes. *Biotechnol. J.*, 4, 1168–1177, 2009.
11. Dhatterwal, J.S., Kaswan, K.S., Jaglan, V., Vij, A., Machine learning and deep learning algorithms for IoD, in: *The Internet of Drones*, pp. 237–292, Apple Academic Press, New Jersey and Canada, 2022.
12. Ji, C., Mandania, R., Liu, J., Liret, A., Kern, M., Incorporating risk in field services operational planning process, in: *International Conference on Innovative Techniques and Applications of Artificial Intelligence, Lecture Notes in Artificial Intelligence*, vol. 11311, Springer, Cham, Switzerland, Cambridge, UK, 2018.
13. Kaswan, K.S., Dhatterwal, J.S., Baliyan, A., Jain, V., Special sensors for autonomous navigation systems in crops investigation system, in: *Virtual and Augmented Reality for Automobile Industry: Innovation Vision and Applications*, pp. 65–86, Springer International Publishing, Cham, 2022.
14. Gao, Q., He, Y., Fu, J.Z., Liu, A., Ma, L., Yang, G., Microfluidic generation of alginate microfibers with tunable diameters and wall thicknesses. *Biomicrofluidics*, 9, 1, 014106, 2015.
15. Xu, C., Chai, W., Huang, Y., Markwald, R.R., Scaffold, L.C., Freeform fabrication of hydrogel scaffolds with tailored biomolecular surfaces. *Biomaterials*, 33, 34, 8657–8668, 2012.
16. Jang, J., Park, J.Y., Gao, G., Cho, D.W., Biomaterials-based 3D cell printing for next-generation therapeutics and diagnostics. *Biomaterials*, 156, 88–106, 2018.
17. Fedorovich, N.E., Alblas, J., De Wijn, J.R., Hennink, W.E., Verbout, A.B.J., Dhert, W.J.A., Hydrogels as extracellular matrices for skeletal tissue engineering: State-of-the-art and novel application in organ printing. *Tissue Eng.*, *13*, 1905–1925, 2007.
18. Bhattacharjee, T., Zehnder, S.M., Rowe, K.G., Jain, S., Nixon, R.M., Sawyer, W.G., Angelini, T.E., Writing in the granular gel medium. *Sci. Adv.*, 4, 7, eaat1637, 2018.
19. Kaswan, K.S., Dhatterwal, J.S., Preeti, K., Intelligent agent based case base reasoning system build knowledge representation in COVID-19 analysis of recovery infectious paitents, in: *Application of AI in COVID 19*, 2020.
20. Ozbolat, I.T. and Hospodiuk, M., Current advances and future perspectives in extrusion-based bioprinting. *Biomaterials*, 76, 321–343, 2016.

21. Rutz, A.L., Hyland, K.E., Jakus, A.E., Burghardt, W.R., Shah, R.N., A multimaterial bioink method for 3D printing tunable, cell-compatible hydrogels. *Adv. Mater.*, 27, 1607–1614, 2015.
22. Dhatterwal, J.S., Kaswan, K.S., Pandey, A., Implementation and deployment of 5G-drone setups, in: *The Internet of Drones*, pp. 47–64, Apple Academic Press, Cham, Switzerland, Cambridge, UK, Cham, 2022.
23. Zhang, A. and Liebscher, D., A review on the computational methods for 3D bioprinting. *Comput. Biol. Med.*, 108, 193–207, 2019.
24. Mandrycky, C., Wang, Z., Kim, K., Kim, D.H., 3D bioprinting for engineering complex tissues. *Biotechnol. Adv.*, 34, 422–434, 2016.
25. Datta, P., Ayan, B., Ozbolat, I.T., Bioprinting for vascular and vascularized tissue biofabrication. *Acta Biomater.*, 51, 1–20, 2017.
26. Li, J., Wu, C., Chu, P.K., Gelinsky, M., 3D printing of hydrogels: Rational design strategies and emerging biomedical applications. *Mater. Sci. Eng. R Rep.*, 140, 100543, 2020.
27. Gao, G., Lee, J.H., Jang, J., Applications of 3D bioprinting in tissue engineering and regenerative medicine. *Bioengineering*, 5, 4, 55, 2018.
29. Miri, A.K., Mirzaee, I., Hassan, S., Recent advances in machine learning approaches for biofabrication and bioprinting. *Mater. Today Bio*, 5, 100040, 2020.
30. Wang, Y., Kim, M.H., Wang, M., Gao, Y., Won, S.M., Cruz-Gonzalez, D., Rogers, J.A., A bioinspired flexible organic artificial afferent nerve. *Sci. Adv.*, 6, 14, eaaz0999, 2020.
31. Wang, M.O., Piard, C.M., Melchiorri, A.J., Dreher, M.L., Fisher, J.P., Mikos, A.G., Tailoring the degradation kinetics of three-dimensional-printed poly (ε-caprolactone) tissue engineering scaffolds. *Tissue Eng. Part A*, 24, 7-8, 582–593, 2018.
32. Pati, F., Jang, J., Ha, D.H., Won Kim, S., Rhie, J.W., Shim, J.H., Cho, D.W., Printing three-dimensional tissue analogues with decellularized extracellular matrix bioink. *Nat. Commun.*, 5, 1, 1–12, 2014.

# Index

Printed and bound by CPI Group (UK) Ltd, Croydon, CR0 4YY
17/09/2026
14964280-0002